Electronic Logic Circuits

J. R. Gibson

Department of Electrical Engineering and Electronics
University of Liverpool.

Edward Arnold

© J. R. Gibson 1979

First published 1979
by Edward Arnold (Publishers) Ltd.
41 Bedford Square
London WC1B 3DQ
Reprinted with solutions to selected problems 1980

ISBN: 0 7131 3407 0

Filmset and printed by Thomson Litho Ltd., East Kilbride, Scotland.

Preface

The introduction of integrated circuits in the mid-1960s and their subsequent development has greatly extended the applications of electronic logic circuits. They are no longer used only in computers and sophisticated instruments but are incorporated in a wide range of products. This requires that students of many branches of engineering, not just electronic engineering, must acquire some knowledge of the capabilities, operation, and design of logic systems.

These rapid changes in technology require corresponding changes in the teaching of logic circuit design. It is no longer a specialist subject taught only to electronic engineering students in their final year of study but is a subject encountered by many students at the start of an engineering course. This book is based on courses given to first- and second-year undergraduate students in the Department of Electrical Engineering and Electronics at Liverpool University. The courses are intended as introductory ones for those students who later specialize in digital electronics and also provide a sufficient background in the subject for those who will only be indirectly involved with logic systems.

An elementary approach is adopted, as most first-year students have little or no relevant background knowledge. Several topics are omitted and others are introduced in terms of a single application to illustrate basic ideas. For example, basic logic functions are introduced directly rather than through set theory and Venn diagrams. This has the advantage that it is not necessary to understand elements of set theory before attempting to understand logic functions. The direct approach to the Boolean functions AND, OR and NOT causes no difficulties when it follows an explanation of the engineering advantages of two-state components. In general students appear to have few difficulties until they encounter sequential circuits; to avoid many of the difficulties that then arise, a single reliable – if lengthy – approach is used for sequential circuit design.

Emphasis is placed on the construction of circuits using elements which are readily available in integrated circuit form. Rather more detail than is customary in a textbook on logic design is given regarding the use, interconnection and limitations of these elements, but details of their internal construction are avoided. Because this practical approach is adopted, the symbols used for logic elements are those used by component manufacturers in their published databooks, design manuals, etc. These symbols do not conform to the current British Standard (BS 3939, section 21; revised July 1977) as most integrated circuits are manufactured by American-based companies.

Problems are included at the end of most chapters since any design technique can only be completely learnt by practice in its use. In addition it is useful for students to construct and test circuits built using standard components. Students who attend the courses which form the basis for this book undertake at least ten hours associated laboratory work.

I wish to thank my colleagues, students, former students and many others for discussions and questions with which they unknowingly assisted in the preparation of this book. Thanks are also due to Motorola Semiconductors Inc., for their help. Finally, the assistance of my wife, Kathleen, in the preparation of the manuscript and its subsequent careful typing by Mary Ballin are appreciated.

J.R. Gibson
1978

Contents

1 Two-state Systems

If some condition of a manufactured article may change or be changed then the article may be classified either as a discrete-state or as a continuous-state system. The term 'system' simply describes the article as a whole and may be applied to a component used in some large assembly or to a complete assembly. A system is described as a continuous-state system when the variable condition, i.e. the state of the system, is able to take any value between certain limits. For instance, the sound output from a radio receiver may be adjusted by the listener to any level from inaudible to the maximum output possible. Similarly, a car driver can select any engine speed required by use of the accelerator control. Both the radio receiver and the car engine are examples of continuous-state systems; such systems are also known as continuously-variable ones.

In contrast to the wide range of states (conditions) possible in a continuous-state system, a conventional electric light switch allows only on or off settings of the light; similarly, a car driver can use the gear lever to select one of the small number of gear ratios available. Both the lighting system with an on–off switch and the car gearbox are examples of discrete-state systems; in such systems only a finite, usually fixed, number of different states are allowed.

Many situations arise in which a designer has to choose to use either a discrete-state or a continuous-state system because it is possible to solve a particular engineering problem by using either type of system. For instance, an on–off light switch may be replaced by a dimmer control which allows the user to select any light level over a wide range. Because designers frequently have to choose to use either a discrete-state or a continuous system they should be aware of the particular advantages and disadvantages of each type of system.

A general comparison for every possible situation cannot be made, but in a large proportion of cases the continuous system is more expensive to manufacture and is less reliable than a discrete-state alternative. For example, the cost of a dimmer light control is typically five times that of an on–off switch. However, the dimmer control has the advantage that the user may select any level of lighting rather than just fully on or off; this wider range of choice given to the user is a common advantage of a continuous system.

1.1 Two-state systems

Discrete-state systems which have only two possible states, for example a simple on–off light switch, are particularly easy to manufacture in many cases. This book is concerned only with systems which are constructed entirely from discrete-state components which are restricted to two possible states, i.e. the components are themselves **two-state systems**. Surprisingly, this very strict limitation placed on the components imposes only one restriction on the complete system; it must be a discrete-state system, but it may have any number of states because the two-state components may be combined together so that they form a multiple-state system.

Two-state components are often called **logic elements** and complete systems which are constructed from such components are **logic systems, logic circuits** or **logic networks**; the reasons for these names should become apparent in Chapter 2. An on–off light switch is a two-state device as are many other types of switch, and an alternative name for logic circuits is **switching circuits**.

Nearly all the circuits which will be described can be constructed using any two-state device as the basic component, but in this book emphasis is placed on the use of electronic two-state components. Electronic logic elements have many advantages when compared with other types; they are exceedingly small, require very little power, operate quickly, and are extremely cheap. A single logic element within an electronic logic system may cost less than 0.1p; manufacturing techniques and component design appear to be continuing to advance rapidly and it is probable that costs will continue to fall.

An additional advantage of modern electronic components is that those used in two-state systems are so reliable that systems containing several hundred million components will operate at high speed for long periods without any failures. If the components were not electronic (e.g. mechanical springs and levers or hydraulic devices) failures would occur so often in such large systems that they would rarely, if ever, operate successfully.

1.2   Electronic two-state systems

There are many ways in which electronic logic circuits may be constructed so that they can have only two states. The most common arrangement is such that any circuit input or output which is above some chosen voltage level is defined to be in one of the two states. When an input or output is below some much lower level it is in the other state. It is essential that there is a well-defined gap between the two voltage levels, and all the logic circuits (and those connected to them) must be designed so that all the inputs and outputs can only take up the specified levels.

It is convenient to give names to these two states so that they are easily identified. If one works entirely with such names all the logic networks designed may be constructed using any type of logic element. In one scheme the upper level is called 1 (ONE) and the lower level is called 0 (ZERO); alternative pairs of names in common use include HIGH and LOW, UP and DOWN, ON and OFF, and TRUE and FALSE. The names 1 and 0 are those which will be used most frequently here but the others will be used when they are particularly suited to some application.

There are many ways to design electronic circuits so that they have this two-state behaviour. Detailed consideration of such designs are to be found in books concerned with electronic design and circuit analysis; any reader interested in such details should refer to one of the many books on these subjects. This book is concerned with the design and behaviour of circuits which are constructed from components, each of which is a complete two-state system which is normally – but not necessarily – an electronic circuit.

1.3   Common electronic logic components

With the exception of a few special applications, most modern electronic logic networks are constructed from two-state components purchased in the form of integrated circuits. An integrated circuit is a complete electronic circuit fabricated on a single piece of pure silicon (which is often called a chip) that is typically about 3 mm square and 1 mm thick. The chip is encapsulated in a protective casing of plastic or ceramic which is much larger than the chip; metal leads are fixed to the chip and extend beyond the plastic or ceramic case, and allow electrical connections to be made to the circuit inputs, outputs and power supply points. This packaging arrangement enables the complete integrated circuit to be easily handled and connected into a larger network by hand or automatic machines. Figure 1.1 is an outline drawing of one of the most common encapsulations, the dual in line (DIL) package.

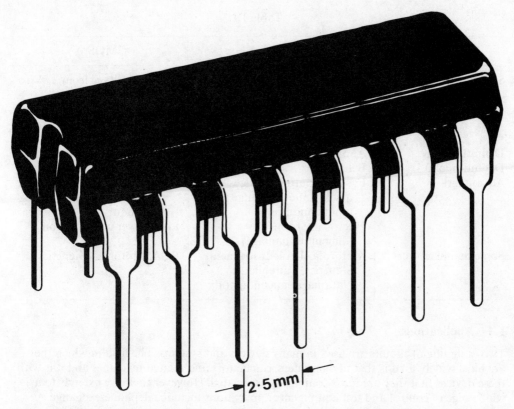

Fig. 1.1 DIL packaged integrated circuit

At present two types of electronic logic components are readily available; transistor–transistor logic (TTL) and complementary symmetry metal oxide semiconductor (CMOS or COSMOS). The two systems are entirely different and should not be interconnected unless special components (interface circuits) are used to make the connections between the two different systems. These two systems supercede several earlier ones, for example resistor–transistor logic (RTL) and diode–transistor logic (DTL).

There is such a large investment in the manufacture of TTL and CMOS components, and in equipment which incorporates them, that these types will probably be the ones most commonly used for some time. One possible future replacement for TTL and CMOS components in some applications may be integrated injection logic (I^2L) components. For applications requiring very high speed of operation (above 100 million operations per second) components using emitter–coupled logic (ECL) are used.

Table 1.1 compares a few features of TTL and COSMOS components; these sytems are now standard ones and devices from different manufacturers are identical in performance.

Except in Chapter 7, no assumptions are made concerning the type of two-state components required; in most cases any type – including non-electronic ones – may be used. Most students of logic systems will find that the construction and testing of logic circuits is an essential part of their studies. For this purpose '7400' series TTL components are the most useful as they are easily obtained and are difficult to damage by incorrect handling or connection. In addition to a small selection of components, some form of prototyping system is required; there are several available at a reasonable cost.

Table 1.1

Property	TTL	CMOS
Supply voltage	5·0V ± 0·25V	Any d.c. voltage from 3·0V to 15·0V
Power consumption	Several milliwatts per element	A function of operating conditions, typically 50 microwatts per element
Definition of 1	Above 2·0V	Above 70% of supply voltage
Definition of 0	Below 0.8V	Below 30% of supply voltage
Advantages	Fast (the output changes quickly after a change in input conditions)	Large voltage fluctuations are required to cause spurious operation
	Electrically robust (difficult to damage)	Low power consumption
Some manufacturers	Motorola, Texas Instruments, Signetics, Fairchild, National Semiconductor	RCA, Motorola, Signetics

1.4   Applications

Two-state (logic) circuits are used in many devices and systems. The field in which they are most widely used is that of computers, calculators and related machines and it is with these devices that they are most commonly associated. However their use extends to a wide range of control and test equipment; applications include telephone-exchange equipment, railway signalling, lift (elevator) controls, etc. Increasingly, some of the complex electronic systems which are now available are being used in domestic products; the most modern automatic clothes washing machines and sewing machines incorporate them and investigations are in progress to examine the advantages and disadvantages of using them in other products.

The dominant position of number processing machines (i.e. computers etc.) as the major application of logic circuits has greatly influenced the development of components and a knowledge of a few of the basic concepts which are required to understand such machines is very useful. In particular, some knowledge of numbers and of coding is required when developing logic-circuit design techniques. This knowledge also assists in understanding some of the literature supplied by component manufacturers.

1.5   Numbers

To understand how a system constructed entirely from two-state components may be used to manipulate numbers it is necessary to examine the principles of number systems. Most people are trained when they are children to use numbers and to perform arithmetic operations but, quite reasonably, very few people are taught the basic theory of numbers and arithmetic.

It is customary to use the decimal number system and to anyone familiar with the system the printed form of a number such as 2093.54 has an exact and easily understood meaning. This printed form is a shorthand representation of

$$2 \times 1000 + 0 \times 100 + 9 \times 10 + 3 \times 1 + 5 \times 0.1 + 4 \times 0.01,$$

which can also be written as

$$2 \times 10^3 + 0 \times 10^2 + 9 \times 10^1 + 3 \times 10^0 + 5 \times 10^{-1} + 4 \times 10^{-2}.$$

This form is just one example of a much more general form of number which is

$$d_n \times b^n + d_{n-1} \times b^{n-1} + \cdots + d_2 \times b^2 + d_1 \times b^1 + d_0 \times b^0 + d_{-1} \times b^{-1} + d_{-2} \times b^{-2} + \cdots$$
$$+ d_{-m} \times b^{-m}$$

where b is the **base** (or radix) of the number system and the ds are the **digits**. In any number system which has this form, the number of different digits required is equal to the base of the system, i.e. the number of different symbols for digits is equal to the base. The origin of the decimal system which has a base of ten is obvious; simple counting and arithmetic is often performed using fingers (digits) to keep a record; each finger therefore represents a different symbol in the number system and a base of ten is a natural consequence.

An advantage of this type of number system is that only a few different symbols (different digits) are required and this small number of symbols may be used to represent any number no matter how large or how small it is. This is possible because the position of any digit relative to the decimal point is important; a digit l places to the left of the point is multiplied by the base to the power $l-1$, i.e. by b^{l-1} and a digit r places to the right is multiplied by the base to the power $-r$, i.e. by b^{-r}. The digits themselves, usually 0, 1, 2, 3, etc., are just printed symbols; they provide one particular method of indicating the quantities zero, one, two and so on. There is no reason why the symbols must be these printed ones; provided that a clearly defined and consistent method is developed, any type of system could be used to represent a digit. One possible scheme is to use the different conditions (states) of a discrete-state system to represent the digits.

When the digits of some number system are represented by the states of a discrete-state system, each state will be used to represent a different digit and the system must have as many states as there are digits in the number system. Therefore, when a two-state system is used to represent digits there can only be two different digits and the number system must have a base of two. The digits in such a system are usually called 0 (zero) and 1 (one) as they must represent nothing and a single unit respectively. These digits may be represented by the two states of a logic element and the reason for calling the states of such an element 0 and 1 is now obvious. The number system which has a base of two is called the **binary system** and most electronic computing devices operate in the binary system or in a modified form of it. A few details of the binary system, including simple methods of conversion between the binary and decimal systems, are given in Appendix A.

Printed binary numbers seem strange when they are first encountered, but once understood they are easier to manipulate than decimal numbers. The most obvious feature of binary numbers – other than the fact that they consist entirely of ones and zeros – is the large number of digits that are required to represent quite small quantities; for example the binary number 101101110 is equivalent to the decimal number 366.

1.6 Codes and coding

Although the binary system is the one which is usually used to represent numbers associated with logic systems, it is not essential to use this number system. Logic elements may be combined to form multiple-state systems, and the states of such systems may be used to represent numbers in systems other than the binary one. When multiple-state systems are built from two-state components the state of each component may be considered to represent one **binary digit** which is usually called a **bit**. If all the bits are considered simultaneously as a binary number with several digits, then each different number represents one state of the multiple-state system. For example, when a system includes three bits denoted by A, B and C there are eight possible different states; some of these states are given in Table 1.2.

Table 1.2

C	B	A
0	0	0
0	1	0
1	1	0
1	0	1
1	1	1

The bits may be considered to be a code in which each combination (i.e. each row in Table 1.2) represents one of the states of a multiple-state system. If the bits of some multiple-state system are required to represent the ten decimal digits then a large number of different codes are possible. A convenient code is one in which four bits A, B, C and D are treated as the digits of a binary number with A as the least significant (2^0) digit and D as the most significant (2^3) digit. Table 1.3 is a listing of this code.

Table 1.3

Bit Weight	D $2^3 = 8$	C $2^2 = 4$	B $2^1 = 2$	A $2^0 = 1$	Decimal digit
	0	0	0	0	0
	0	0	0	1	1
	0	0	1	0	2
	0	0	1	1	3
	0	1	0	0	4
	0	1	0	1	5
	0	1	1	0	6
	0	1	1	1	7
	1	0	0	0	8
	1	0	0	1	9

Each row in Table 1.3 is the binary code for a single decimal digit; because each bit represents a definite power of two the full name of this scheme is 'the 8421 weighted binary coded decimal representation'. This is often shortened to just 'binary coded decimal' or 'BCD' and although many other codes such as 5421 and 2421 weighted binary ones are used to represent decimal numbers, the use of the initials BCD without any qualification implies the 8421 weighted code.

Other codes exist in which there is no weighting, i.e. each bit is not given a specific weight and a look-up approach is normally required to interpret the code. Table 1.4 shows an example of one such code for the digits 0 to 5.

The code in Table 1.4 is a Gray code; it is one of many Gray codes, all of which have the property that the number of bits which are 1 always increases or decreases by one when moving from one position to the next one in the code sequence. Gray codes have the advantage that they give rise to fewer errors when optical or mechanical reading

Table 1.4

C	B	A	Digit
0	0	0	0
0	0	1	1
0	1	1	2
1	1	1	3
1	1	0	4
1	0	0	5

devices are used. Consequently, Gray codes are extensively used in positioning devices in automatic equipment, particularly automatic machine tools.

Coding is a convenient method of indicating the condition or state of a multiple-state system constructed from two-state components. Such coding is frequently used to describe logic circuits but the representation is only occasionally referred to as a code.

1.7 Problems

1 Which of the following contain at least one component which is a discrete-state system?
  (a) A door lock.
  (b) A water tap.
  (c) A pendulum clock.
  (d) An electric clock (alternating current supply).
  (e) A telephone dial mechanism.
  (f) A thermostat (e.g. on an oven or a water heater).
  (g) A controller for traffic signals at a road junction.
  (h) The braking system of a bicycle or car.
2 Read appendix A then perform the following conversions.
  (a) Decimal numbers 357, 68, 59·72 and 23·375 to binary.
  (b) Binary numbers 11010, 1011, 1110111·101 and 101010·011 to decimal.
  (c) Octal (base eight) number 276 to decimal.
  (d) Binary number 11101110 to octal.
3 Devise a four-bit binary code to represent the ten decimal digits. The code must be such that if it is interpreted as a binary number this number would have a value three greater than the decimal digit represented by the code. (This is called the excess three code.)
4 What is the minimum number of binary digits (bits) required to represent a single octal (base eight) digit? Devise a code for all the duodecimal (base twelve) digits using the smallest number of bits necessary.

2 Basic Elements of Combinational Logic

Systems constructed so that all the inputs and outputs can only take either one of two allowed states were introduced in Chapter 1 and termed logic systems or circuits. If each output of such a system depends only on the present states of the inputs to the circuit it is called a **combinational logic circuit.** In a combinational system there is no dependence of one output on the other outputs. Also, input states which have occurred previously have no influence on the circuit behaviour. An alternative way of stating this last fact is that the order in which the inputs are applied to the circuit does not affect its final output. Such a system is shown schematically in Fig. 2.1.

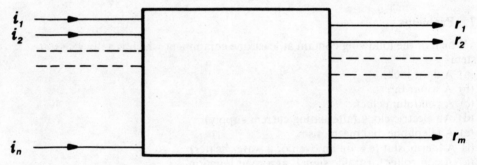

Fig. 2.1. Combinational logic network

In the general case illustrated by Fig. 2.1 there are n inputs $i_1, i_2, i_3, \ldots, i_n$ and m outputs or results $r_1, r_2, r_3, \ldots, r_m$; each input may take either one of the two logic states and the outputs are also restricted to these logic levels. The actual level at a particular output at any time depends on the logic states present at all of the inputs at that instant in time.

It is assumed that as soon as any input changes then the outputs change immediately to the levels which correspond to the new input conditions. Any real circuit will take a finite time to operate; this time is called the propagation delay and it may be neglected in most simple applications of combinational logic circuits.

2.1 Truth tables

Instead of attempting to examine the general case of a system with a large number of outputs, it is sufficient to consider the case of a system with several inputs and a single output, R. The general case of a circuit with m outputs is just m separate single output circuits which all have the same inputs connected to them; i.e. all the circuits have the same n inputs but each has a different output.

When a circuit has n inputs 2^n different input situations may arise because each input may be 0 or 1 and all possible combinations of input states may exist. For example, if there are three inputs A, B and C, then there are eight different input situations; all eight

Table 2.1

C	B	A
0	0	0
0	0	1
0	1	0
0	1	1
1	0	0
1	0	1
1	1	0
1	1	1

are shown in Table 2.1. In this table there is a separate column for every input and each row corresponds to one of the possible combinations of input states.

For each input situation the output, R, must take the value 0 or 1. An n input circuit has 2^n possible combinations of input states and for each of these R must be either 0 or 1; hence $(2^n)^2 = (2^2)^n = 4^n$ different combinational logic circuits exist which have n inputs and a single output. Table 2.2 shows all sixteen possible circuits which have only two inputs; each of the columns headed R_1, R_2,..., etc., corresponds to the output of a different circuit.

Table 2.2

Inputs		Possible Outputs															
A	B	R_1	R_2	R_3	R_4	R_5	R_6	R_7	R_8	R_9	R_{10}	R_{11}	R_{12}	R_{13}	R_{14}	R_{15}	R_{16}
0	0	0	0	0	0	0	0	0	0	1	1	1	1	1	1	1	1
0	1	0	0	0	0	1	1	1	1	0	0	0	0	1	1	1	1
1	0	0	0	1	1	0	0	1	1	0	0	1	1	0	0	1	1
1	1	0	1	0	1	0	1	0	1	0	1	0	1	0	1	0	1

Obviously a circuit with two inputs and a single output behaves so that its output corresponds to one of the output columns of Table 2.2. The behaviour of any single-output combinational logic circuit may be described by a table which is similar to Table 2.2 but has only one output column. Table 2.3 is an example of one possible table for a circuit with three inputs.

Table 2.3 is called a **truth table** and it completely specifies the behaviour of a particular logic circuit. The rules for constructing a truth table are that there must be separate columns for each input and each output (as indicated previously, circuits may have more than one output but can be divided into separate single-output circuits); the table must have one row for each possible combination of input states; every possible input combination must be included and the output(s) produced by the circuit must be shown in all cases.

A truth table is one of the most useful ways of describing the behaviour of a combinational logic circuit and one should be included as a part of the specification of any circuit.

Table 2.3

Inputs			Output
C	B	A	R
0	0	0	0
0	0	1	1
0	1	0	1
0	1	1	0
1	0	0	1
1	0	1	0
1	1	0	0
1	1	1	0

2.2   A set of basic logic elements

It was stated earlier that $(2^n)^2$ different logic circuits may be specified which have n inputs and a single output. Each one of these circuits has its own unique truth table. If n is 2 there are sixteen possible circuits; when n is 3 there are sixty-four; when n is 4 there are two hundred and fifty-six and so on. The number increases rapidly with n and it would be an uneconomic proposition to manufacture different circuits for every possible case. Fortunately, it is possible to choose a set consisting of a small number of basic logic elements selected so that several may be interconnected to produce a circuit which behaves in the manner required by some specification.

There is no unique set of basic logic elements but the five which follow are probably the most useful ones; in addition they directly implement a set of mathematical rules which may be used to describe the behaviour of logical systems.

The basic logic elements used in combinational logic systems are usually called 'gates' by electronic engineers. This is because one common application is to control the flow of electrical logic signals along some path; the element may allow the signal to pass or may prevent its passage and this action is analagous to opening or closing a gate across a path.

2.2.1   The AND gate

This element has *a single output and any number of inputs.* It is defined as the combinational logic circuit which gives the output 0 unless all the inputs are at the logic 1

Table 2.4

Inputs			Output
C	B	A	R
0	0	0	0
0	0	1	0
0	1	0	0
0	1	1	0
1	0	0	0
1	0	1	0
1	1	0	0
1	1	1	1

level; in this case only the output is 1. This definition is sufficiently precise to allow the truth table of an AND gate to be written down immediately. Table 2.4 is the truth table for a three-input AND gate.

When this element is used as a component in a large system it is convenient to be able to represent it by some quickly recognized symbol, so that circuit diagrams of the network may be easily drawn and understood. Figure 2.2 shows several different symbols which are used to represent an AND gate. In all cases the inputs are from the left (three inputs are shown) and the output is towards the right.

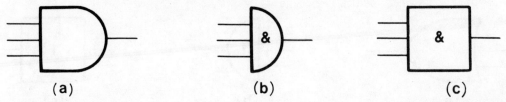

(a)   **(b)**   **(c)**

Fig. 2.2. AND gate symbols

The rectangular symbol of Fig. 2.2c is now the accepted British Standard, but most integrated circuit and electronic equipment manufacturers still use the symbol of Fig. 2.2a. Designers must refer to the data which is published by manufacturers in order to use components correctly. To avoid the necessity of students having to learn both sets of symbols while still unfamiliar with logic elements and systems, the manufacturers' practice (Fig. 2.2a) is adopted here. This is also the practice of most journals concerned with electronic circuit design and manufacture.

One further method which is used to describe any logic circuit is to express its behaviour in terms of Boolean algebra. This algebra is the mathematical description of systems which are only allowed to have two states and it will be examined in more detail later. An essential definition in Boolean algebra is the AND function, and in equations the function is represented by a dot. For example, the Boolean expression for a three-input AND gate is the equation

$$R = A \,.\, B \,.\, C$$

which is read as R *equals* A *and* B *and* C.

2.2.2 The OR gate

This element is another one which implements a fundamental definition in Boolean algebra. The OR gate is the combinational logic circuit which has any number of inputs

Table 2.5

Inputs			Output
C	B	A	R
0	0	0	0
0	0	1	1
0	1	0	1
0	1	1	1
1	0	0	1
1	0	1	1
1	1	0	1
1	1	1	1

and has an output of 1 when any one, or more than one, of the inputs is 1. An alternative statement is that the circuit gives the output 1 unless all the inputs are 0, in which case it gives the output 0. Note that a two-input OR circuit is not just 1 in the cases input A is 1 with B equals 0, or input B is 1 with A equals 0; it also includes the situation where A and B are both 1 (i.e. A is 1 or B is 1 or A and B are both 1). From this description the truth table can be constructed; Table 2.5 is that for a three-input OR gate.

Figure 2.3 shows some of the symbols which are used to represent an OR gate; Fig. 2.3c is the new standard but Fig. 2.3a is the most commonly used one.

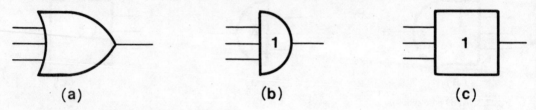

(a)   **(b)**   **(c)**

Fig. 2.3. OR gate symbols

In Boolean algebra the OR function is represented by a plus sign and the equation describing a three-input example is

$$R = A + B + C$$

which is read as R *equals* A *or* B *or* C.

2.2.3   The INVERTER or NOT gate

Unlike the two gates described previously this gate is only defined in the case in which it has a single input and a single output whereas the others may have any number of inputs. This is the one gate which is essential in any basic set of logic functions. Its action is to produce an output logic state which is not the same as the state at the input. Because only two states are possible in a logic system this statement is a complete description. The truth table of an inverter is obvious and is shown in Table 2.6.

Table 2.6

Input A	Output R
0	1
1	0

Many different symbols are used to represent this gate; several of them are shown by Fig. 2.4. Again, the rectangular symbol is the standard one for this gate, but the triangular symbol is the most commonly used. The method of representing inversion in Boolean algebra is to draw a line above the quantity to be inverted. Therefore, if R is the result of inverting A, the equation which describes this operation is

$$R = \bar{A}$$

which is read as R *is not* A. The inverse of a logical quantity is termed its **complement**; in the equation above, R is the complement of A.

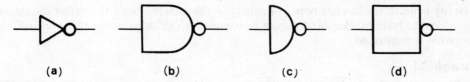

(a)   (b)   (c)   (d)

Fig. 2.4. INVERTER symbols

2.2.4 The NAND gate

This gate is equivalent to an AND gate followed by an inverter (NOT gate). However, the convention in Boolean algebra is to state the NOT before the quantity which is to be inverted so this gate is said to perform the operation NOT–AND; this is usually contracted to NAND. Despite this slightly confusing name, the output of a NAND gate must be the exact inverse of the output of an AND gate; i.e. a NAND gate will always give the output 1 unless all the inputs are 1, in which case only the output is 0. Table 2.7 is the truth table of a two-input NAND gate.

Table 2.7

Inputs		Output
B	A	R
0	0	1
0	1	1
1	0	1
1	1	0

This gate may be represented by the symbol for an AND gate followed by that for an inverter, as in Fig. 2.5a, but because the NAND gate is used so often the symbol is contracted to the form shown in Fig. 2.5b. The small circle after the AND symbol indicates inversion; this convention of a small circle following another symbol to indicate inversion is used in most pictorial representations of logic elements. For example it is used for the two other NAND gate symbols shown in Figs 2.5c and 2.5d.

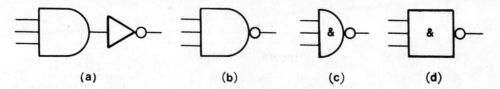

(a)   (b)   (c)   (d)

Fig. 2.5. NAND gate symbols

In terms of Boolean algebra this is written as

$$R = \overline{A.B}$$

which is R equals not the result of A and B, or alternatively, R *equals NAND of A with B*. The statement 'R equals not A and B' is not clear – it could mean that R is the quantity $\overline{A}.B$ or that it is the quantity $\overline{A.B}$.

When a line is drawn above a quantity the inversion must not be performed until the quantity beneath the line has been evaluated; i.e. the quantity beneath the line should be considered to be in brackets and should be completely evaluated before the inversion operation is performed.

Example 2.1

The quantity $R = \overline{A.B.C}$ is $R = \overline{(A.B.C)}$ and this is not the same as $X = \bar{A}.\bar{B}.\bar{C}$ which is $X = (\bar{A}).(\bar{B}).(\bar{C})$. Prove that the quantities X and R are different.

Solution

Table 2.8

Inputs			Intermediates				Outputs	
C	B	A	$\bar{C}$	$\bar{B}$	$\bar{A}$	$Y = A.B.C$	$R = \bar{Y}$	$X = (\bar{A}).(\bar{B}).(\bar{C})$
0	0	0	1	1	1	0	1	1
0	0	1	1	1	0	0	1	0
0	1	0	1	0	1	0	1	0
0	1	1	1	0	0	0	1	0
1	0	0	0	1	1	0	1	0
1	0	1	0	1	0	0	1	0
1	1	0	0	0	1	0	1	0
1	1	1	0	0	0	1	0	0

A truth table is written out in full for the quantities X and R which are compared. This method is sometimes called perfect induction. It is useful to define an intermediate quantity $Y = A.B.C$; and Table 2.8 is the result. Examination of the columns formed for R and X shows that they are different, i.e. $\overline{A.B.C}$ is not the same as $\bar{A}.\bar{B}.\bar{C}$.

2.2.5   The NOR gate

This gate is equivalent to an OR gate followed by an inverter; i.e. the NOR function is NOT–OR. The NOR gate has an output of 0 unless all the inputs are 0, in which case the output is 1. The behaviour of a two-input NOR gate is given in Table 2.9. In Boolean algebra this is written as $R = \overline{A+B}$ which is R equals not the result of A or B, or alternatively, R *equals NOR of* A *with* B.

Table 2.9

Inputs		Output
B	A	R
0	0	1
0	1	0
1	0	0
1	1	0

The OR symbol can be modified to show inversion, just as the AND symbol was, by drawing a small circle after the symbol for OR to produce the symbol for a NOR gate. Several NOR gate symbols are shown in Fig. 2.6.

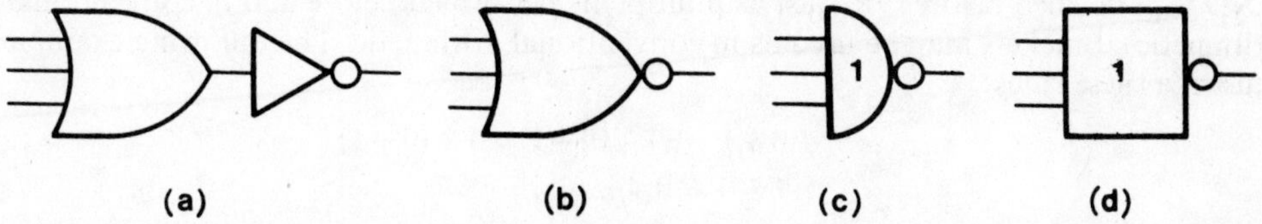

Fig. 2.6. NOR gate symbols

As in the case of the NAND function, care is required in stating the inversion operation; for example, $\overline{A+B+C}$ is not the same as $\bar{A}+\bar{B}+\bar{C}$ and the difference can be shown by writing out complete truth tables for both expressions.

Note that the truth tables for a single-input NAND gate and a single-input NOR gate are identical and are the same as the truth table for an inverter. It is quite common for designers of large logic systems to connect all the inputs of a multiple-input NAND or NOR gate together (it then becomes a single-input gate) to provide an inverter. This construction is common when it simplifies the layout of a circuit. It is usually used because integrated circuits are often supplied with two or more multiple-input gates in the same package; if some of the gates in a package are not required in a particular application it may be economical to use them as inverters.

2.3 Boolean algebra

Boolean algebra is a development of the work of the nineteenth century mathematician George Boole. It is used in the same way as conventional algebra to equate and manipulate variable quantities which are represented by symbols (e.g. A,B,C, etc.). However, in Boolean algebra the variables are only allowed to take either one of two states called true and false (or 1 and 0) and the rules for the manipulation of these variables are not the same as the rules used in conventional algebra.

Conventional algebra is based on standard arithmetic with the four operations of add, subtract, multiply and divide. When Boolean variables are used these operations no longer apply; the operations of Boolean arithmetic must be used and they are the logical operations AND, OR, and NOT which have already been described. As there are only two values of a Boolean variable the rules of Boolean arithmetic are easily summarized and are illustrated by the examples in Table 2.10.

Table 2.10

$$\left.\begin{array}{l}\bar{1}=0\\\bar{0}=1\end{array}\right\} \text{NOT}$$

$$\left.\begin{array}{l}0.0=0\\0.1=1.0=0\\1.1=1\end{array}\right\} \text{AND (for two variables)}$$

$$\left.\begin{array}{l}0+0=0\\0+1=1+0=1\\1+1=1\end{array}\right\} \text{OR (for two variables)}$$

These rules are easily extended to more than two variables as the normal associative, commutative and distributive laws of conventional algebra hold. In other words, the order of performing operations does not matter except that in an expression without any brackets

AND is performed before OR (just as multiply is performed before add in conventional arithmetic). Brackets may be used as in conventional arithmetic. The following examples illustrate these rules:

$$1+0+1 = (1+0)+1 = 1+(0+1)$$
$$1+0 = 0+1$$
$$1.1+0.1 = (1.1)+(0.1)$$
$$1.0.1 = (1.0).1 = 1.(0.1)$$

Standard arithmetic operations lead to conventional algebra and these Boolean operations lead to Boolean algebra. Table 2.11 is a list of examples of the relationships which hold in Boolean algebra, together with some comments. Many relationships appear

Table 2.11

SOME RELATIONSHIPS IN BOOLEAN ALGEBRA

$\left.\begin{array}{l} A.1 = A \\ A.0 = 0 \\ A+1 = 1 \\ A+0 = A \end{array}\right\}$	Zero and unit rules. (Note dominance of 0 in AND and 1 in OR)
$\left.\begin{array}{l} A.\bar{A} = 0 \\ A+\bar{A} = 1 \\ \bar{\bar{A}} = A \end{array}\right\}$	Complement relations
$\left.\begin{array}{l} A+A = A \\ A.A = A \end{array}\right\}$	Idempotence
$\left.\begin{array}{l} A+B = B+A \\ A.B = B.A \end{array}\right\}$	Commutative laws
$\left.\begin{array}{l} A+A.B = A \\ A.(A+B) = A \\ A+\bar{A}.B = A+B \end{array}\right\}$	Absorption rules
$\left.\begin{array}{l} A.(B+C) = A.B+A.C \\ A+B.C = (A+B).(A+C) \end{array}\right\}$	Distributive laws
$\left.\begin{array}{l} A+B+C = (A+B)+C = A+(B+C) \\ A.B.C = A.(B.C) = (A.B).C \end{array}\right\}$	Associative laws (insertion of brackets)
$\left.\begin{array}{l} \overline{A+B+C+D+\cdots} = \bar{A}.\bar{B}.\bar{C}.\bar{D}.\cdots \\ \overline{A.B.C.D.\cdots} = \bar{A}+\bar{B}+\bar{C}+\bar{D}+\cdots \end{array}\right\}$	de Morgan's Theorem

to be exactly the same as those of conventional algebra but others seem unusual when first encountered; e.g. $A.A = A$ and $A.(A+B) = A$. Although most of the examples in Table 2.11 have a maximum of three variables, the rules may be applied to any number of variables. For instance

$$A.B+A.C+A.D.E = A.(B+C+D.E)$$

All the relationships may be shown to be correct by deriving the complete truth tables for both sides of the identity.

Example 2.2

Verify de Morgan's theorem for three variables in the form

$$\overline{A.B.C} = \bar{A} + \bar{B} + \bar{C}$$

Solution

Table 2.12

C	B	A	$W = A.B.C$	$\overline{W} = \overline{A.B.C}$	$Z = \bar{C}$	$Y = \bar{B}$	$X = \bar{A}$	$X + Y + Z = \bar{A} + \bar{B} + \bar{C}$
0	0	0	0	1	1	1	1	1
0	0	1	0	1	1	1	0	1
0	1	0	0	1	1	0	1	1
0	1	1	0	1	1	0	0	1
1	0	0	0	1	0	1	1	1
1	0	1	0	1	0	1	0	1
1	1	0	0	1	0	0	1	1
1	1	1	1	0	0	0	0	0

Define intermediate quantities of $W = A.B.C$, $X = \bar{A}$, $Y = \bar{B}$ and $Z = \bar{C}$ and then write out a complete truth table for both sides of the identity; this is Table 2.12.

The columns for $\overline{A.B.C}$ and $\bar{A} + \bar{B} + \bar{C}$ are identical and show that de Morgan's theorem holds in this case.

These relationships in Boolean algebra may be used to manipulate Boolean expressions (that is mathematical descriptions of logic circuits) into alternative forms. The usual reason for performing such manipulations is to change one Boolean expression for a circuit into another which is more easily or more economically constructed from the available components.

Example 2.3

Suppose the output, R, of some circuit is given as

$$R = A.C.D + A.\bar{B}.\bar{C}.D + A.B.\bar{C}.D + \bar{A}.\bar{B}.\bar{C}.\bar{D}$$

Can this be manipulated into a more simple form?

Solution

Change the expression to

$$R = A.C.D + A.\bar{C}.D.(\bar{B} + B) + \bar{A}.\bar{B}.\bar{C}.\bar{D}$$

Since $B + \bar{B} = 1$ then it can be written as

$$R = A.C.D + A.\bar{C}.D + \bar{A}.\bar{B}.\bar{C}.\bar{D}$$

which can be further treated in the same way giving

$$R = A.D.(C + \bar{C}) + \bar{A}.\bar{B}.\bar{C}.\bar{D}$$
$$= A.D + \bar{A}.\bar{B}.\bar{C}.\bar{D}$$

This final form is much simpler than the original, but some steps in the reduction were not obvious unless the answer was already known.

One fault of Boolean algebra is that the steps in the manipulation are not always obvious. Chapter 3 introduces some methods which may be used to reduce Boolean expressions to simpler forms in a reliable and consistent manner.

2.4   de Morgan's theorem

de Morgan's theorem (or law) is very important; it is probably one of the most frequently used relationships in Boolean algebra. The two forms are

$$\overline{A+B+C+\cdots} = \bar{A}.\bar{B}.\bar{C}.\cdots$$

and

$$\overline{A.B.C.\cdots} = \bar{A}+\bar{B}+\bar{C}\cdots$$

and both forms are valid with any number of variables.

The principle use of de Morgan's theorem is to convert an OR type of expression (i.e. either OR or NOR) into an AND form (i.e. either AND or NAND) and *vice versa* when a particular type of logic gate is to be used for circuit construction. For example, suppose that $X = A+B+C$ and that a circuit is required to produce X using only NAND gates. Because double inversion is the same as no inversion, then $X = \overline{\overline{A+B+C}}$; applying de Morgan's theorem to $\overline{A+B+C}$ the result becomes

$$X = \overline{\bar{A}.\bar{B}.\bar{C}}$$

Remembering that an inverter is a single input NAND gate then the circuit shown in Fig. 2.7 will behave as an OR gate; this circuit is constructed entirely from NAND gates.

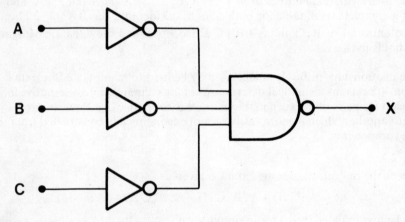

Fig. 2.7. OR function generated by NAND gates

de Morgan's theorem is a mathematical expression which describes the property of all two-state systems which is known as duality.

2.5   Duality

This is a rather difficult concept to grasp and although it is not essential in the design and use of logic circuits, it is a useful concept in the understanding of basic principles. The treatment here is brief and the reader should consult more advanced texts for more detailed information.

Basically the property of duality is that provided all input *and* output logic levels are inverted, AND becomes OR and OR changes to AND. For example, Table 2.13a is the truth table for a two-input AND gate while Table 2.13b is produced from it by changing every 1 to 0 and every 0 to 1.

Examination of Table 2.13b shows that it is the truth table of an OR gate, so the process of inverting all logic levels has changed and AND gate into an OR gate. But initially it was necessary to select which logic state was called 1 and which was 0; the

Table 2.13

Inputs		Output		Inputs		Output
B	A	R		$\bar{B}$	$\bar{A}$	$\bar{R}$
0	0	0		1	1	1
0	1	0		1	0	1
1	0	0		0	1	1
1	1	1		0	0	0
(a)				(b)		

choice was entirely arbitrary and there is no reason why the opposite definition cannot be made and used. Hence, when the logic levels are defined in one way a particular circuit may behave as an AND gate, but when they are defined in the opposite way *the same circuit* is an OR gate.

2.6 The exclusive-OR function

A basic set of logic elements has been selected; in fact the set is too large because it is possible to construct every possible logic circuit using only NAND gates or only NOR gates (provided an inverter is regarded as a single input NAND gate or a single input NOR gate). A larger range of elements is retained because circuit designs may be more economical when a range is available.

One circuit is required so often that it is convenient to regard it as another logic element. When the OR function was introduced it was emphasized that a two-input OR gate has an output of 1 if either input is at 1 or if both are at 1. An alternative function would be one which gives an output of 1 when either input is 1 but *not* when both inputs are 1. Such a circuit is called an **exclusive-OR** gate (because it excludes the case that both inputs are 1) and unlike the normal OR gate it is only defined as a circuit with two inputs. The truth table for the **exclusive-OR** gate is shown as Table 2.14.

Table 2.14

Inputs		Output
B	A	R
0	0	0
0	1	1
1	0	1
1	1	0

The exclusive-OR function is often indicated by the symbol $\oplus$ in equations. Hence,

$$R = A \oplus B.$$

Although the exclusive-OR function is only defined for two inputs, an equation such as $X = A \oplus B \oplus C \oplus D$ is exact because the same result is obtained whichever order the exclusive-OR operations are performed in. That is $(A \oplus B) \oplus (C \oplus D)$ gives exactly the same result as $((A \oplus B) \oplus C) \oplus D$, but these multiple-input systems are not called exclusive-OR gates; this name is only applied to the two-input gate.

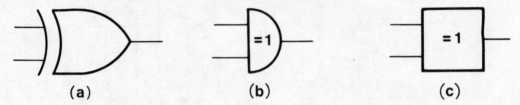

(a)   **(b)**   **(c)**

Fig. 2.8. EXCLUSIVE-OR gate symbols

Some of the circuit symbols for exclusive-OR gates are shown in Fig. 2.8. The exclusive-OR function is available as an integrated circuit but it is also easily constructed from the standard functions. For example, it can be shown that using NAND gates

$$R = A \oplus B = \overline{(A.\bar{B}).(\bar{A}.B)} = \overline{(\overline{(A.B)}.A).(\overline{(A.B)}.B)}$$

and these NAND gate circuits are shown in Fig. 2.9.

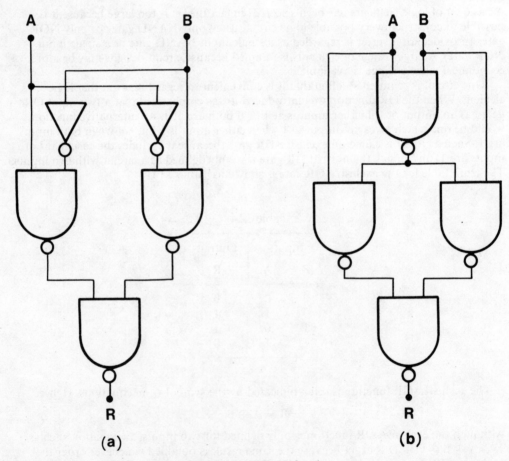

(a)   **(b)**

Fig. 2.9. NAND gate implementations of the EXCLUSIVE-OR function

2.7 Logic circuit analysis

Those readers who are familiar with linear electronic circuit analysis will be aware that the analysis process is difficult. In the case of combinational logic circuits analysis is a trivial (although tedious) task. The aim of any circuit analysis is to start from a circuit diagram and to obtain an exact description of the behaviour of the circuit; when the circuit is a combinational logic one, the result of an analysis will be either a Boolean expression or a truth table.

To perform the analysis it is sufficient to work systematically through the circuit from the inputs to the outputs, determining the output of every gate.

Example 2.4

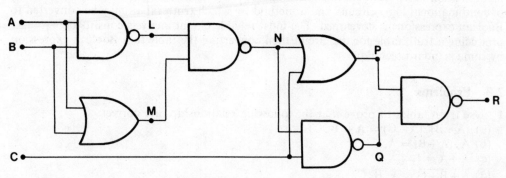

Fig. 2.10.

Analyse the circuit behaviour of the combinational logic system shown in Fig. 2.10.

Solution

All the gate outputs have been labelled; it is just necessary to write down a truth table step by step in the way that is indicated by Table 2.15.

Table 2.15

C	B	A	$L = \overline{A.B}$	$M = A + B$	$N = \overline{L.M}$	$P = N + C$	$Q = \overline{N.C}$	$R = \overline{P.Q}$
0	0	0	1	0	1	1	1	0
0	0	1	1	1	0	0	1	1
0	1	0	1	1	0	0	1	1
0	1	1	0	1	1	1	1	0
1	0	0	1	0	1	1	0	1
1	0	1	1	1	0	1	1	0
1	1	0	1	1	0	1	1	0
1	1	1	0	1	1	1	0	1

This technique is tedious but it will always give the correct result. An alternative method is to work in terms of Boolean algebra; i.e. from the circuit diagram,

$$L = \overline{A.B} \text{ and } M = A + B.$$

Also, $$N = \overline{L.M}.$$

Hence, $$N = \overline{L.M} = \overline{(\overline{A.B}).(A+B)} = \overline{\overline{A.B}} + \overline{(A+B)} = A.B + \bar{A}.\bar{B}.$$

Further, $Q = \overline{N.C}$ and $P = N+C$ so that since $R = \overline{Q.P}$, the same manipulations that were used with $N = \overline{L.M}$ give

$$R = N.C + \bar{N}.\bar{C}.$$

Hence, $N.C = (A.B + \bar{A}.\bar{B}).C = A.B.C + \bar{A}.\bar{B}.C.$

Also, $\bar{N}.\bar{C} = \overline{(A.B + \bar{A}.\bar{B})}.\bar{C} = A.\bar{B}.\bar{C} + \bar{A}.B.\bar{C}.$

Therefore the circuit behaviour is described by

$$R = A.B.C + \bar{A}.\bar{B}.C + A.\bar{B}.\bar{C} + \bar{A}.B.\bar{C}.$$

The Boolean solution is not always obvious and it is easy to make errors in the manipulation of Boolean expressions. Chapter 3 is concerned with the design (synthesis) of combinational logic circuits and a method by which truth tables may be converted to Boolean expressions is developed. The most reliable circuit analysis consists of producing a truth table for the circuit then converting the table to a Boolean expression by some standard technique.

2.8   Problems

1   Use truth tables to prove that the following relationships are correct:
(a) $(A+B).(A+C) = A + (B.C)$,
(b) $A.(A+B) = A$,
(c) $A + \bar{A} = 1$,
(d) $\overline{A+B+C} = \bar{A}.\bar{B}.\bar{C}$,
(e) any relationships in Table 2.11 which do not appear to be reasonable ones.

2   Simplify the following Boolean expressions:
(a) $A.\bar{B}.\bar{C} + A.B.\bar{C} + \bar{A}.\bar{C}$,
(b) $M.\bar{N}.P + \bar{L}.M.P + \bar{L}.M.\bar{N} + \bar{L}.M.N.\bar{P} + \bar{L}.\bar{N}.\bar{P}$,
(c) $A.B.\bar{C}.\bar{D} + \bar{A}.B.\bar{D} + \bar{A}.\bar{B}.C + \bar{B}.D$.

3   Show that the expression $R = \overline{(A.\bar{B})}.\overline{(\bar{A}.B)}$ is equivalent to the exclusive-OR function.

4   Prove that the following expressions involving the exclusive-OR function are correct:
(a) $A \oplus B = \bar{A} \oplus \bar{B}$,
(b) $\overline{A \oplus B} = \bar{A} \oplus B = A \oplus \bar{B}$.
Note that expression (b) is sometimes called an exclusive-NOR function.

5   Derive the truth table and a Boolean expression for the circuit shown in Fig. 2.11.

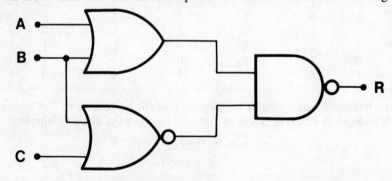

Fig. 2.11.

3 The Design of Combinational Logic Circuits

When a designer receives the specification for a combinational logic circuit it may be a rather vague written or verbal description of the task to be performed by the circuit. Alternatively, the specification may be a complete truth table for the circuit, or it may be somewhere between these two extremes. Before any attempt can be made to produce a detailed circuit design an exact description of the circuit behaviour is required; the most useful description is a complete truth table for the circuit.

3.1 Development of a truth table

The first stage in any systematic approach to the design of a combinational logic circuit is the production of the truth table for the circuit. However, since the initial specification may be given in many different ways, no general rules can be formulated for the construction of the truth table. In common with many design situations, much of the process must be left to the experience and common sense of the designer.

To produce a truth table, the number of circuit inputs and outputs must be determined in those cases for which they are not explicitly specified. A table of all the possible combinations of input states can then be constructed. Table 2.1 is one example for a circuit with three inputs. This table becomes the truth table when the state of each output is added to the table alongside every input combination. In some cases the specification will impose constraints which influence the form of the truth table to a great extent; in other cases the designer will have a large element of choice in forming the table.

Example 3.1

Devise a truth table for a circuit with the following behaviour. Three simple two-position switches are connected to supply inputs to a logic circuit which has a single output. The output controls a lamp and the circuit operation is such that changing the position of any switch changes the condition of the lamp (i.e. if the lamp was on it goes off and *vice versa*.)

Solution

This specification is such that the number of circuit inputs and outputs are determined but many other features must be selected by the designer.
(a) Inputs. Arbitrarily choose one switch position, call it down, and let it represent an input of 0; the other position, up, then represents an input of 1. Also distinguish the three inputs by labelling them A, B and C.
(b) Output. Again a free choice of logic states exists; it would seem sensible to choose an output of 1 to cause the lamp to be on and an output of 0 to correspond to the lamp off.
(c) Starting position. Although the specification states how the circuit must change when any input changes, it does not indicate the output for any input condition. In this example it is necessary to choose the output for a single set of inputs. A reasonable choice is to decide that if all the inputs are 0, then the output is also 0.

Once these decisions have been made for this particular example the complete truth table can be constructed. Starting with all three inputs at 0 the output is also 0. The

Table 3.1

Inputs			Output
C	B	A	
0	0	0	0
0	0	1	1
0	1	0	1
0	1	1	0
1	0	0	1
1	0	1	0
1	1	0	0
1	1	1	1

specification requires that a change at a single input will change the output. Hence the three different cases with one input of 1 and the other two inputs 0 must all be cases in which the circuit will give an output of 1. A similar argument shows that the three cases with two inputs of 1 and the third 0 must be cases for which the output is to be 0. Finally, when all three inputs are 1, the output must be 1 because this is a single switch changed from the two 1s and one 0 case; the complete truth table, Table 3.1, can now be formed.

Example 3.2

Obtain the truth table for the circuit which has three inputs A, B and C and one output R. The circuit behaviour is given by the Boolean relationship

$$R = A.(B + \bar{B}.\bar{C}) + \bar{A}.\bar{B}.C$$

Solution

This is another possible form of circuit specification and it is one which leaves very few decisions to be made by the designer. The numbers of inputs and outputs are determined by the specification and the truth table is obtained simply by evaluating the Boolean expression for every possible set of input conditions. This evaluation is made simple when intermediate quantities (as in Example 2.1) are defined; choose these to be $L = \bar{B}.\bar{C}$, $M = B + L = B + \bar{B}.\bar{C}$, $N = A.M = A.(B + \bar{B}.\bar{C})$ and $P = \bar{A}.\bar{B}.C$. When R is evaluated for all combinations of A, B and C, Table 3.2 is produced.

Table 3.2

Inputs			Intermediates				Output
C	B	A	$L = \bar{B}.\bar{C}$	$M = B + L$	$N = A.M$	$P = \bar{A}.\bar{B}.C$	$R = N + P$
0	0	0	1	1	0	0	0
0	0	1	1	1	1	0	1
0	1	0	0	1	0	0	0
0	1	1	0	1	1	0	1
1	0	0	0	0	0	1	1
1	0	1	0	0	0	0	0
1	1	0	0	1	0	0	0
1	1	1	0	1	1	0	1

3.2   Minterms and maxterms

It is not possible to devise a single formal process which can be used to develop a truth table from all forms of circuit specification, but when the truth table has been prepared it is possible to devise systematic methods of circuit design. These methods require the truth table as the starting-point and a single method is introduced in this chapter; it is one of the most commonly used design techniques.

The first stage in this circuit-design procedure is to derive a Boolean expression which describes the circuit behaviour. This may appear to be unnecessary if, as in Example 3.2, the original specification is itself a Boolean expression for the circuit. However, the form of expression in a specification will probably contain an arbitrary mixture and AND, OR and inversion operations; in order to approach the design of combinational logic circuits in a systematic manner a specific form of Boolean expression is required. Therefore, the truth table is not just to be converted into a Boolean expression, but is to be expressed in one with a particular form.

Because logic expressions have a dual nature (see Chapter 2) every technique developed can be formulated in two different ways. Therefore, when a method of deriving a Boolean expression from a truth table is devised, there will also be a related dual method. Both methods will lead to the same result, but in the case of the method introduced here, one system is slightly easier to use; this is the technique which is based on minterms.

A **minterm** is that AND function which includes the algebraic symbols for every input to the circuit (usually called input variables, variables or literals) once, and only once, in either true or complemented (inverted) form. Furthermore, each minterm corresponds to a single row in the truth table in such a way that the minterm can only have the value 1 when the circuit inputs have the values corresponding to those for that particular row of the truth table. That is, each minterm is uniquely associated with a single row in the truth table and may be used to describe that row. Table 3.3 shows all the possible combinations of inputs for a system which has three inputs of A, B and C, together with the minterms which correspond to each case.

Table 3.3

Inputs			
C	B	A	Minterm
0	0	0	$\bar{A}.\bar{B}.\bar{C}$
0	0	1	$A.\bar{B}.\bar{C}$
0	1	0	$\bar{A}.B.\bar{C}$
0	1	1	$A.B.\bar{C}$
1	0	0	$\bar{A}.\bar{B}.C$
1	0	1	$A.\bar{B}.C$
1	1	0	$\bar{A}.B.C$
1	1	1	$A.B.C$

The alternative, or dual, expression is called a **maxterm** and is an OR function which includes each input variable once only in either true or complemented form.
Unfortunately, there are two different definitions of a maxterm in common use; this causes confusion and is one of the reasons why minterms are used here. The possible forms of maxterms and their use to describe circuits are briefly discussed in Appendix B; the results obtained using maxterms can be shown to be the same as those from minterms.

The extension of minterms (and maxterms) to four or more variables is simple. For example, a four-input system will have minterms such as $A . B . C . D$. Because they have a printed form which is similar to an algebraic product, minterms are often called **products** or product terms. Similarly, maxterms may be referred to as **sums** or sum terms.

3.3   Minterm representation of circuits

The conversion of a truth table into a Boolean expression which consists entirely of minterms is straightforward. Each minterm uniquely represents a single row of the truth table, it can only have the value 1 when all the inputs have the values which correspond to those in the row of the truth table associated with it. If all the minterms which correspond to those conditions for which the output must be 1 are listed, and only these minterms are listed, then if input conditions exist for which the circuit is required to give an output of 1, a single one of the listed minterms will have the value 1. In those cases for which the circuit output is to be 0 all the listed minterms will be 0. The OR function is defined such that it produces the result 1 when any single input (or more than one input) has the value 1. Therefore, if all the listed minterms are combined by an OR function, the complete expression produced is one which exactly describes the behaviour of the circuit represented by the truth table.

Writing the truth table for Example 3.2 (Table 3.2) again, without the intermediate columns but with a column of minterms, Table 3.4 is obtained. The minterms with an asterisk are those which correspond to cases for which the final circuit must produce an output of 1.

Table 3.4

Inputs			Output	Minterm
C	B	A	R	
0	0	0	0	$\bar{A} . \bar{B} . \bar{C}$
0	0	1	1	$A . \bar{B} . \bar{C}$*
0	1	0	0	$\bar{A} . B . \bar{C}$
0	1	1	1	$A . B . \bar{C}$*
1	0	0	1	$\bar{A} . \bar{B} . C$*
1	0	1	0	$A . \bar{B} . C$
1	1	0	0	$\bar{A} . B . C$
1	1	1	1	$A . B . C$*

In this example, the output must have a value of 1 if any one of the minterms $A . \bar{B} . \bar{C}$, $A . B . \bar{C}$, $\bar{A} . \bar{B} . C$ and $A . B . C$ is 1, therefore, the output, R, is given by

$$R = A . \bar{B} . \bar{C} + A . B . \bar{C} + \bar{A} . \bar{B} . C + A . B . C.$$

The similarity between this form of expression and one in conventional algebra is such that this is often called a **sum of products**. After the truth table has been developed for a combinational logic circuit, the next step in the systematic design of the circuit is to produce this minterm sum of products expression.

3.4   Minimization

A circuit could be constructed immediately from the sum of products expression by using inverters, AND gates, and one multiple-input OR gate. Such a circuit would operate correctly, but would usually be more expensive to manufacture than one in which the Boolean expression had been simplified in some way. In the example above, the minterm formulation gave the relationship

$$R = A.\bar{B}.\bar{C} + A.B.\bar{C} + \bar{A}.\bar{B}.C + A.B.C.$$

Using the rules of Boolean algebra, this may be reduced to

$$R = A.B + A.\bar{C} + \bar{A}.\bar{B}.C.$$

This second form obviously requires fewer logic elements, and furthermore, most of the elements required for this reduced form have fewer inputs than those in the original. It is reasonable to assume that the smaller the number of inputs to a logic element the less complicated – and consequently the cheaper – it will be.

The usual aim of any engineer is to produce a design which completely meets the specification of the system required at the lowest possible cost. Both expressions for R clearly meet the circuit specification but it is probable that the one with the smaller number of elements (and also with elements which have fewer inputs) will be the cheaper one. An additional advantage of the reduced form of circuit is that the smaller the number of components used in any circuit, the lower the chance of a component failure when the circuit is in use. Thus, the more simple form is probably the more reliable one, and it should be cheaper to maintain.

To achieve minimum cost, the circuit designer has to solve a minimization problem. It is assumed here (and in most other treatments of logic circuit design) that the cheapest circuit is the one which requires the smallest number of elements, and minimal cost is regarded as equivalent to producing the minimum form of Boolean expression for the circuit. In a few special cases the minimum form of logic expression is not the cheapest one to implement, but in such cases the minimum expression is rarely much more expensive than the cheapest form.

After producing the minterm form of Boolean expression for a circuit, the next stage of circuit design is to reduce this expression to a minimum form. It was indicated in Chapter 2 that the reduction of Boolean expressions by inspection, using the rules of Boolean algebra, is not obvious in many cases. Some method is required which will assist with this reduction; when five or fewer variables are involved, one of the most useful techniques is that which uses a Karnaugh map.

3.5 Karnaugh maps

Karnaugh maps are a modification of Venn diagrams which are a pictorial device that should be familiar to any reader with some elementary knowledge of set theory. It is not essential to understand the origins of Karnaugh maps in order to use them to simplify

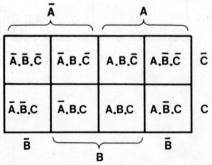
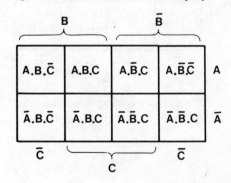

Fig. 3.1. Karnaugh maps

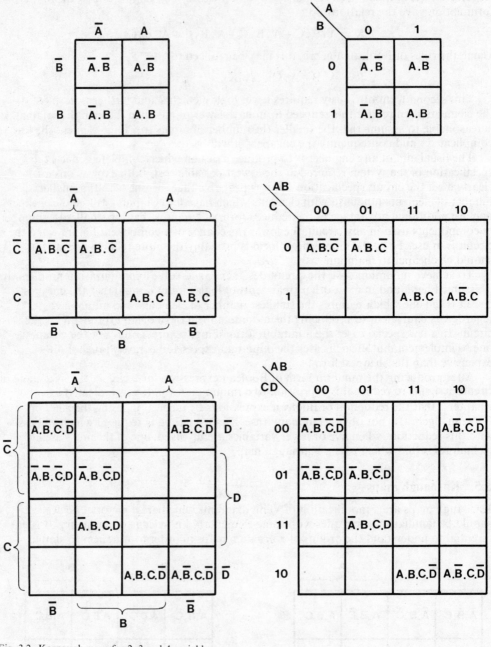

Fig. 3.2. Karnaugh maps for 2, 3 and 4 variables

logical expressions; some of the more advanced texts listed in the bibliography include details of the basic theory of the technique.

A Karnaugh map consists of a rectangular area which is divided into squares (or elements) and each square represents one minterm. There is only one square for any minterm and there is a square for every minterm. The squares are not allocated to the minterms at random, but are arranged so that a movement of one square vertically (up or down) or one square horizontally (left or right) results in the minterms associated with the two adjacent squares differing only in a single variable. In other words, the two minterms are identical except for one variable which is inverted in one of them but not in the other. Diagonal movements on the Karnaugh map are not of interest. Within these rules many different maps may be drawn. Figure 3.1 shows two different maps for a system with three inputs of A, B and C; both maps are correct.

A Karnaugh map is best regarded as a three-dimensional device which has to be represented in a two-dimensional form when it is printed. This is one reason why so many maps may be drawn. When movements of one square are repeatedly made in the same direction, the edge of the map is eventually reached. A further move of one square is equivalent to moving off the edge of the map and returning onto it at the opposite edge; i.e. the top edge should be joined to the bottom one and the left-hand edge should be joined to the right-hand one. The only way in which the map could be drawn with this form is on a doughnut-shaped surface (the surface of a toroid). The two-dimensional map is an attempt to represent this surface and it is satisfactory provided that it is remembered that a movement off one edge of the map and back on at the opposite edge is identical to a move from one square to an adjacent square.

Figure 3.2 shows examples of 2-, 3- and 4-variable maps with two different methods of labelling each map. In most cases only a few of the individual squares are labelled with minterms; the form of labelling of each row and column enables the minterm for any square to be determined. Some readers will find one of the forms of labelling easier to use than the other. Usually, the form on the right-hand side of Fig. 3.2 is easier to use when inserting logic values from a truth table into the map, while the form on the left of Fig. 3.2 is better when the completed map is used for logic expression reduction. Figure 3.3 is one form of 5-variable map and consists of two 4-variable maps which should be imagined to be placed one above the other.

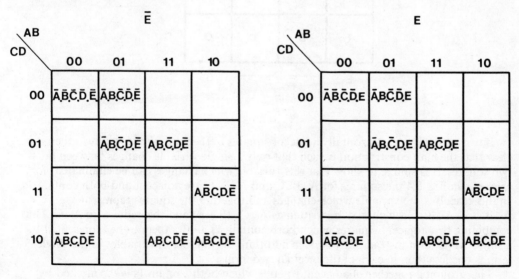

Fig. 3.3. Karnaugh map for 5 variables

For systems which have more than five variables, the Karnaugh map technique is not very successful and alternative methods are required. (With care, 6-variable maps consisting of four 4-variable maps placed one above the other can be used.) However, a very large number of circuits involve five or fewer variables, or may be reduced to several such circuits, so methods of logic expression minimization based on Karnaugh maps may be applied to many circuit design problems. In cases for which the maps cannot be used other methods, for example those developed by Quine and McCluskey, are required and are explained in more advanced texts.

3.6   Minimization using Karnaugh maps

To produce a minimal form of Boolean expression with the aid of a Karnaugh map the map must be completed by entering either 1 or 0 in *every* square of the map. Each square corresponds to one minterm (i.e. to one row of the truth table) and 1 is put in those squares which correspond to the cases for which the circuit output is required to be 1, 0 is entered when the output is to be 0. If a circuit has several outputs a separate map should be drawn for each output.

With practice a Karnaugh map may be completed directly from the truth table, but initially the sum of products expression should be formed and used to complete the map.

In Section 3.3 the expression

$$R = A.\bar{B}.\bar{C} + A.B.\bar{C} + \bar{A}.\bar{B}.C + A.B.C$$

was obtained for Example 3.2. The Karnaugh map for R is completed by writing 1 in each square corresponding to the minterms in the expression for R and writing 0 in all the other squares; the completed map for R is shown in Fig. 3.4.

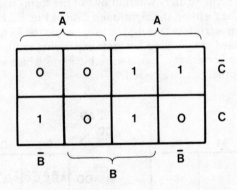

Fig. 3.4.

If two vertically or horizontally adjacent squares in the map both contain 1s then it is clear that the map construction is such that only a single variable changes between the two squares. It is simple to show that this variable which changes can be eliminated; for example, in Fig. 3.4 the squares for $A.B.\bar{C}$ and $A.\bar{B}.\bar{C}$ are adjacent and both contain 1s. In this case, B is the variable which changes and together the squares represent $A.B.\bar{C} + A.\bar{B}.\bar{C}$ which may be written as $A.\bar{C}.(B + \bar{B})$ and this reduces to $A.\bar{C}$. Thus, combining two adjacent squares which both contain 1s allows them to be represented by a single AND term instead of by two; in addition, the number of variables in the single term is one less than in each of the original two terms.

This grouping together of adjacent squares which both contain 1s is the method by which a sum of products expression is simplified with the aid of a Karnaugh map. The technique can be extended by grouping together adjacent, but not overlapping, groups of two squares to form larger groups. These larger groups *must* be square or rectangular and must have sides which are 2^a squares by 2^b squares where a and b are zero or positive integers. Therefore, on a four-variable map, the groups may be a single square, two squares (2 by 1), four squares (either 2 by 2 or 4 by 1), eight squares (4 by 2) or the complete map. Five- and six-variable maps may have three-dimensional groups of 2^a by 2^b by 2^c where c is also zero or integral.

Each group of 1s formed is equivalent to replacing all the minterms represented by the squares in the group by a single AND expression which contains only those variables which are the same in every square in the group. Thus, all those variables which change are eliminated, and the larger the group, the greater the degree of simplification; if the whole map contains 1s then all the variables change and the circuit output is 1, regardless of the input conditions.

To produce a minimum logic expression using a Karnaugh map all the 1s must be included in groups which are 2^a by 2^b and the groups must be as large as possible. When constuctiong the groups *it is essential to remember that the maps are global*, i.e. a movement off one edge and back on to the map at the opposite edge is exactly the same as a movement between adjacent squares. Figure 3.5 illustrates a range of different groups on some four variable maps; some of the groups which run off the edges are not obvious.

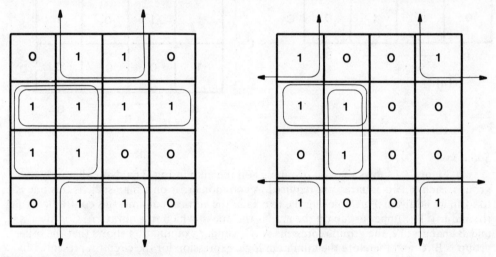

Fig. 3.5. Formation of groups on Karnaugh maps

When groups are formed, every 1 must be included in at least one group but may be included in more than one group if this allows a greater reduction of the whole minterm expression. In other words, it is not essential to include a 1 in every possible group which could be formed to include it, but when it is already in one group it may be included in another if this allows a small group to be replaced by a larger one. For the greatest degree of minimization to be achieved, the smallest possible number of groups should be formed.

The following rules summarize the method which should be adopted to produce a minimum logic expression with a Karnaugh map.

    (a) Form the *largest* possible groups.

    (b) Construct the smallest possible number of groups provided that rule (c) is obeyed.

    (c) All the squares which contain a 1 must be included in at least one group.

    (d) Squares should not be included in more than one group *unless* the inclusion of a square in more than one group enables a small group to be replaced by a larger one.

Inexperienced designers commonly make the mistakes of using too many groups or constructing groups which are too small. The second fault is most frequently found when groups run off the edges of the map or when groups overlap. Mistakes such as these do not produce a faulty circuit; the circuit operates correctly but it is not the simplest one possible.

Example 3.3

The Karnaugh maps in Fig. 3.6 illustrate the required behaviour of two logic circuits. Derive minimum logic expressions for both circuits.

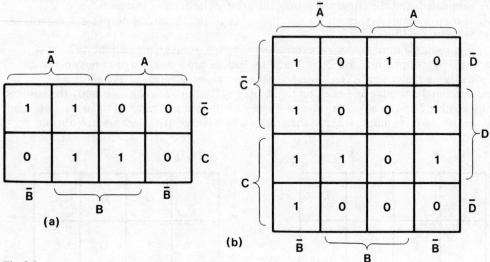

Fig. 3.6.

Solution

(a) Figure 3.7 shows the best groupings on the map in Fig. 3.6a. Note that only two groups, each of two squares, are required. A common error on a map such as this one is to form an additional group using the two 1s in the vertical column; this contradicts rules (b) and (d). The upper group on the map is the one in which A is always $\bar{A}$, C is always $\bar{C}$, and B changes; i.e. the group represents $\bar{A}.\bar{C}$. Similar examination shows that the other group is $B.C$ and therefore the minimum logic expression for the circuit represented by the map is $\bar{A}.\bar{C}+B.C$.

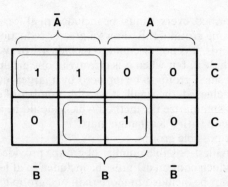

Fig. 3.7.

(b) Figure 3.8 indicates the groups for the map of Fig. 3.6b; in this scheme the group
$\bar{A}.\bar{B}.C.D$ has been included in three different groups. If the column of four squares is
considered to be an obvious first group, then the remaining four 1s must be included in
additional groups. The two 1s on the right-hand edge could be taken as a two-by-one
group; however, if the two 1s on the opposite edge which have already been used in the

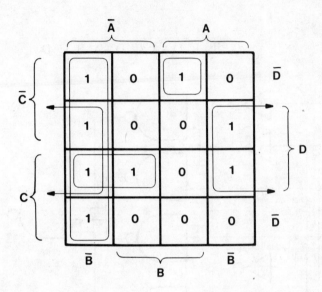

Fig. 3.8.

column group are also included, a larger two-by-two group is obtained. Similarly the
square $\bar{A}.B.C.D$ can be grouped with the square $\bar{A}.\bar{B}.C.D$ which has already been
used twice; now only the square $A.B.\bar{C}.\bar{D}$ remains to be included but as there are no
squares containing a 1 adjacent to it then the only group possible is the single square
itself. The solution consists of four groups: the column $\bar{A}.\bar{B}$; the group $\bar{B}.D$ which runs
off the edges; the group of two $\bar{A}.C.D$; and the single square $A.B.\bar{C}.\bar{D}$. Therefore the
minimized expression for the circuit is

$$\bar{A}.\bar{B}+\bar{B}.D+\bar{A}.C.D+A.B.\bar{C}.\bar{D}$$

3.7 Circuit implementation

The minimum expression which is obtained using the reduction techniques described here
is another sum of products form and the circuit could be constructed using several AND
gates and one OR gate. However, the principal objective of the minimization technique
was the reduction in the cost of the final circuit. If the circuit can be constructed from a
single type of gate then bulk purchase of the gates will probably reduce the cost; also, in
some electronic logic systems, NAND or NOR gates are the most easily constructed types
and are therefore the cheapest.

The lowest cost is usually achieved when a circuit is constructed entirely from NAND
gates or entirely from NOR gates (an inverter is regarded as either a single input NAND
gate or a single input NOR gate). The sum of products expression obtained from a
Karnaugh map minimization is easily converted into either an expression of NAND
functions or one of NOR functions by application of de Morgan's theorem.

Example 3.4

Convert the expression for the output, X, of a circuit into both NAND and NOR function forms. X is the result obtained in part (b) of Example 3.3, i.e.

$$X = \bar{A}.\bar{B} + \bar{B}.D + \bar{A}.C.D + A.B.\bar{C}.\bar{D}$$

Solution

(a) Invert X twice

$$X = \bar{\bar{X}} = \overline{\overline{\bar{A}.\bar{B} + \bar{B}.D + \bar{A}.C.D + A.B.\bar{C}.\bar{D}}}$$

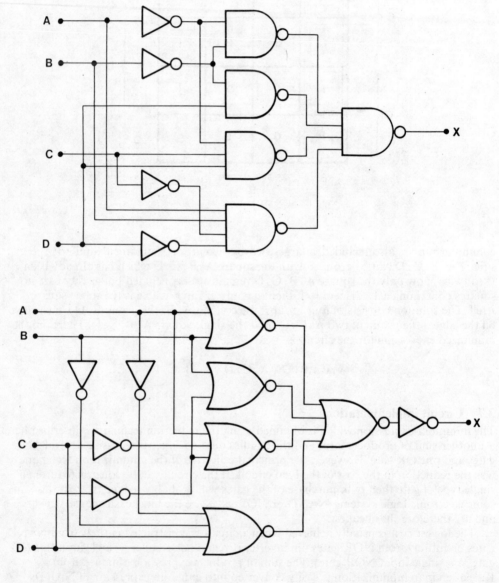

Fig. 3.9. NAND and NOR gate circuits for the function X

Apply de Morgan's theorem to $\bar{X}$, that is to the first inversion. The result is

$$X = \overline{(\bar{A}.\bar{B}).(\bar{B}.D).(\bar{A}.C.D).(A.B.\bar{C}.\bar{D})}$$

which is entirely formed from NAND functions as required.

(b) To obtain the NOR form, each AND term is separately inverted twice giving

$$X = \overline{\overline{\bar{A}.\bar{B}}} + \overline{\overline{\bar{B}.D}} + \overline{\overline{\bar{A}.C.D}} + \overline{\overline{A.B.\bar{C}.\bar{D}}}$$

Again de Morgan's theorem is used and the result is

$$X = \overline{A+B} + \overline{B+\bar{D}} + \overline{A+\bar{C}+\bar{D}} + \overline{\bar{A}+\bar{B}+C+D}$$

Double inversion of this expression gives one which is entirely in NOR functions.

Diagrams for both forms of the circuit are drawn directly from the expressions for X and are shown in Fig. 3.9.

3.8 Unspecified states

In the preceding sections it has been assumed that for every possible input condition the circuit specification explicitly requires a particular output from the circuit. However, in some applications of logic circuits it is impossible for certain input conditions to occur, and in these cases the circuit output will not be specified; the following is a simple example of one such situation.

An industrial process requires that a section of the production plant is kept between two temperatures. If the temperature exceeds some upper limit a refrigerator is switched on, and if the temperature falls below a lower limit a heater is switched on. Two thermostatic switches, A and B, provide an indication of the temperature to a control unit; switch A gives a signal of 1 when the temperature exceeds the upper limit and 0 when it does not, while switch B gives 1 when the temperature is too low and 0 otherwise. The control unit is a combinational logic circuit, the signals from A and B are the inputs, and there are two outputs, R and H. Output R controls the refrigerator so that when $R = 1$ the refrigerator is on, and when $R = 0$ it is off. Similarly, when $H = 1$ the heater is on, and when $H = 0$ it is off. Table 3.5 is the truth table for this logic circuit.

Table 3.5

Inputs		Outputs		
A	B	R	H	Comments
0	0	0	0	Temperature is within required limits
1	0	1	0	Temperature too high, cooling required
0	1	0	1	Temperature too low, heating required
1	1	?	?	Impossible input condition; would imply temperature above upper and below lower limits.

If a Karnaugh map is used to produce a minimized logic expression, how can cases such as $A = B = 1$ in the temperature controller be taken into account? As this condition can never arise, no particular outputs are required. However, to use the Karnaugh map it is necessary to make an entry in every square so the outputs must be chosen to have a value in this case but it does not matter if the value chosen is 1 or 0. Because the circuit operation is not important in this instance these cases are known as **don't care conditions** or just as **'don't cares'**. A second type of 'don't care' situation can arise and will be introduced in Chapter 6; although the origin of the case is different the treatment is identical.

The aim of logic expression minimization was to reduce the circuit complexity and cost. 'Don't cares' can be used to assist in minimization by choosing the circuit outputs in 'don't care' cases to have the values which give the most simple circuit. When 'don't cares' are used in this way they must be indicated in the truth table and on the Karnaugh map by some symbol; the letter X is used here. The minimization technique itself is modified by choosing a 'don't care' condition to give an output of 1 when this reduces the final logic expression and to give an output of 0 in all other cases. The rules for Karnaugh map minimization remain the same with the additional requirement that Xs are to be included in groups of 1s when this increases the group size or reduces the number of groups, but the Xs are not included in groups in other cases. Thus all the 1s must be included in groups but the inclusion of Xs depends on their position relative to 1s.

Note that in the temperature controller example, the case $A = B = 1$ might arise if one detector develops a fault. A third circuit output could be provided to switch on an alarm in this case; this addition would not affect the design of the circuits which provide the other outputs.

Example 3.5

The decimal digits 0 to 9 are represented by an 8421 BCD code (see Section 1.5). Derive a Boolean expression for a logic circuit which will produce an output of 1 when any code representing a digit which is an integral multiple of three is input to the circuit.

Solution

As four Boolean variables with sixteen possible combinations are used to represent the ten decimal digits, six-input combinations will never arise and may be considered to be 'don't care' situations. The truth table is easily constructed and is Table 3.6. The table

Table 3.6

Decimal digit	BCD code				Output R	Minterm
	D	C	B	A		
0	0	0	0	0	1	$\bar{A}.\bar{B}.\bar{C}.\bar{D}$
1	0	0	0	1	0	$A.\bar{B}.\bar{C}.\bar{D}$
2	0	0	1	0	0	$\bar{A}.B.\bar{C}.\bar{D}$
3	0	0	1	1	1	$A.B.\bar{C}.\bar{D}$
4	0	1	0	0	0	$\bar{A}.\bar{B}.C.\bar{D}$
5	0	1	0	1	0	$A.\bar{B}.C.\bar{D}$
6	0	1	1	0	1	$\bar{A}.B.C.\bar{D}$
7	0	1	1	1	0	$A.B.C.\bar{D}$
8	1	0	0	0	0	$\bar{A}.\bar{B}.\bar{C}.D$
9	1	0	0	1	1	$A.\bar{B}.\bar{C}.D$
Unused	1	0	1	0	X	$\bar{A}.B.\bar{C}.D$
,,	1	0	1	1	X	$A.B.\bar{C}.D$
,,	1	1	0	0	X	$\bar{A}.\bar{B}.C.D$
,,	1	1	0	1	X	$A.\bar{B}.C.D$
,,	1	1	1	0	X	$\bar{A}.B.C.D$
,,	1	1	1	1	X	$A.B.C.D$

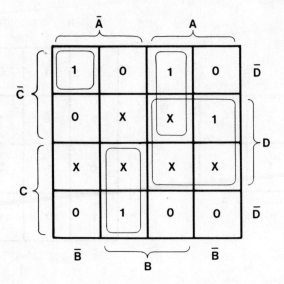

Fig. 3.10.

shows that four minterms correspond to outputs of 1, and six to 'don't care' cases.
Figure 3.10 is the Karnaugh map derived from the truth table with the groups indicated;
these groups correspond to the expression for the output which is

$$R = A.D + A.B.\bar{C} + \bar{A}.B.C + \bar{A}.\bar{B}.\bar{C}.\bar{D}$$

The result in Example 3.5 is a simpler one than that which would have been obtained
if the six unspecified cases had been chosen to give all 1s or all 0s as outputs. In the
groups selected there are four Xs; the effect of including any X in a group is to change the
X into a 1, and any X which is not in a group becomes a 0.

3.9 Summary of the design method

The design procedure described produces a reliable and economical combinational logic
circuit if correctly applied; it may be summarized as follows.
  (a) Obtain a precise circuit specification.
  (b) Convert the specification into a truth table which shows the outputs required for
every combination of input conditions; 'don't cares' are used for the outputs when the
input conditions can never exist.
  (c) Identify the minterms corresponding to each row in the table.
  (d) Draw and complete the Karnaugh map for the system.
  (e) Select groups on the completed maps; form the largest possible groups and the
smallest number of groups.
  (f) Determine the sum of products expression for the selected groups.
  (g) Convert the sum of products expression into the one which is most easily
implemented using the available components.
  (h) Draw the circuit diagram.

3.10 Particular maps

Sometimes the Karnaugh map produced when a circuit is designed is similar to one of the
maps in Fig. 3.11. These maps resemble a chessboard with a 1 in the position of every
black square and a 0 in the position of every white square. Using the usual simplification

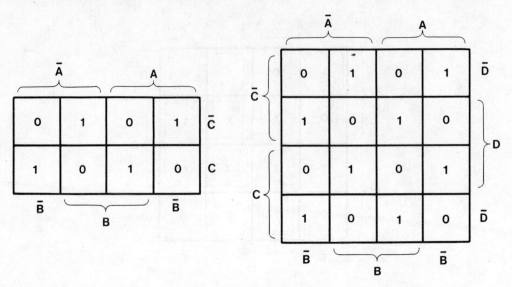

Fig. 3.11. Karnaugh maps for circuits involving EXCLUSIVE-OR gates

rules no groups can be formed on these maps. However, these maps arise when the circuit can be formed very simply using exclusive-OR gates. For example, the map on the left in Fig. 3.11 represents the function $A \oplus B \oplus C = (A \oplus B) \oplus C = A \oplus (B \oplus C)$; this may be proved by deriving the Karnaugh map for the function.

3.11   Multiple-output circuits

It has been assumed that multiple-output circuits are designed as a number of independent single-output circuits. This is reliable, but techniques exist which enable more economical multiple-output designs to be produced. These techniques generally require that the user has some experience of logic circuit design, and a reasonable compromise for the beginner is to design separate single-output circuits and then examine the minimum expressions carefully. If these minimized expressions are inspected for terms common to more than one expression, the common terms need only be generated once within the complete multiple-output circuit.

Example 3.6

Draw the circuit diagram in NAND gate form for a circuit with three inputs A, B and C and three outputs X, Y and Z. The outputs are given by

$$X = A.B.\bar{C} + A.\bar{B}.C$$

$$Y = \bar{A}.B + A.\bar{B}.C$$

$$Z = \bar{A}.B + B.\bar{C}$$

Solution

It is apparent that some of the product terms are common to more than one output expression; if $L = \overline{\bar{A}.B}$ and $M = \overline{A.\bar{B}.C}$ then the outputs are given by

$$X = A.B.\bar{C} + \bar{M}$$

$$Y = \bar{L} + \bar{M}$$

$$Z = \bar{L} + B.\bar{C}$$

Applying de Morgan's theorem gives the NAND forms

$$X = \overline{\overline{(A.B.\bar{C})}.M}$$

$$Y = \overline{L.M}$$

$$Z = \overline{L.\overline{(B.\bar{C})}}$$

L and M may be generated using inverters for $\bar{A}$ and $\bar{B}$ and one NAND gate for each. The outputs of these NAND gates are used as inputs to the circuits which generate the final outputs. The complete circuit is shown in Fig. 3.12. Some degree of economy can be achieved in this way but methods which give greater economy do exist.

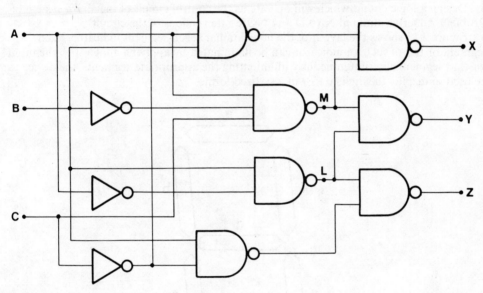

Fig. 3.12

3.12 Comment

A single circuit design technique has been described, it is one of many that have been developed. The technique produces an economical circuit design in nearly all cases, and the circuit is a reliable one which completely meets the initial specification. There are a few exceptional cases in which the minimal solution is not obtained, but even in these cases the circuit is reliable. Many other techniques exist and should be examined by the reader when familiar with this one.

3.13 Problems

1   Derive the truth table for the circuit whose output R is given by

$$R = A.\bar{B}+B.(C+A.D).$$

2   A piece of equipment has four fault detectors and these are connected into an alarm control unit which is a logic circuit. A single fault is regarded as unimportant, it probably arises from a faulty detector. Devise the truth table for the control circuit which indicates the presence of a fault by the state of the output. (Faults are therefore indicated only if two or more inputs – i.e. detectors – show a fault condition.)

3   Use the Karnaugh map technique to reduce the following expressions to a more simple form.

(a) $A.C + A.B.\bar{C} + A$.

(b) $A.B.C + A.C.D + \bar{B}.C.D + \bar{A}.B.\bar{C}.D + \bar{A}.B.\bar{C}.\bar{D}$.

(c) $R.T + \bar{R}.S.T + \bar{R}.\bar{S}.T$.

(d) $L.M.N.P + L.M.\bar{P} + L.\bar{M}.N + \bar{L}.M.\bar{P} + L.N + L.\bar{M}.N.P$.

(e) $\overline{x.\bar{y}.z}.(\bar{x}+y+z) + \overline{x.y}$.

4   Two binary numbers, a and b, each of two digits are represented as $A_1\,A_0$ and $B_1\,B_0$ where A_1, A_0, B_1 and B_0 are Boolean variables which form the digits of the binary numbers. Devise a combinational logic circuit, or circuits, to provide outputs of X, Y and Z such that X = 1 when $a > b$, Y = 1 when $a = b$, and Z = 1 when $a < b$.

5   Design a logic circuit which will operate as the control circuit of the alarm system of Problem 2. Derive minimal NAND and NOR gate versions of the circuit.

6   Figure 3.13 shows the layout of the now familiar seven-segment indicator. Each segment or bar of such an indicator can be illuminated by applying a logic 1 to the input for that segment; by simultaneously illuminating the appropriate segments, the device can be used to display the digits 0 to 9 in a stylized form.

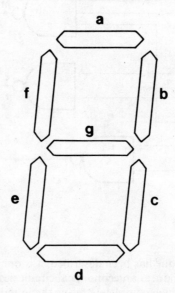

Fig. 3.13. Seven-segment indicator layout

Design a logic circuit, or circuits, whose seven outputs drive a seven-segment display. The circuit has four inputs; the logic levels at these inputs represent the ten decimal digits in an 8421 BCD code.

4 Sequential Logic Elements

The circuits described in the preceding chapters are such that the outputs specified by the truth table appear as soon as the inputs are connected. In any real circuit there will be a small time interval between the connection of inputs and the appearance of the outputs because the circuit takes a finite time to operate. This interval is called propagation delay and has no effect on the final output(s) of the circuits. One important feature of the combinational logic circuits which were examined earlier is that output states previously held by such circuits have no effect on their present behaviour.

When the output of a circuit depends on past inputs (and hence on existing or previous outputs) as well as the present inputs, it is important to consider the circuit action as a function of time. Any logic circuit in which the order in time of applying inputs is important is termed a **sequential circuit**, i.e. the inputs must follow a specific sequence to produce a required output. In order to follow a sequence of inputs the circuit must contain some form of memory to retain knowledge of those inputs which have already occurred. This memory is usually obtained by feedback connections which are made so that the effect of the earlier inputs is maintained.

Sequential logic systems are usually divided into two groups: synchronous and asynchronous circuits. A synchronous circuit or system is one in which all the changes take place simultaneously at a time determined by a signal at some control input common to all sections. In an asynchronous system there is no common control; a change in one section of the system causes further changes in other sections and so on. The changes propagate through the system in a manner which is determined only by the speed with which each section operates.

Most sequential systems are based on a small number of simple sequential circuit elements known as **bistables** or **flip-flops**, so-called because they have two stable conditions and can be switched from one to the other by appropriate inputs. These stable conditions are usually called the **states** of the circuit.

4.1 The set–reset flip-flop

This is the most simple sequential circuit element and is commonly referred to by its initials as an SR flip-flop. The circuit may be constructed in many ways; Fig. 4.1a is one form using NAND gates and Fig. 4.1b is another using NOR gates.

The circuit has two inputs, S and R, and two outputs, Q and Q'. The feedback mechanism required to form a sequential circuit can be clearly seen; each of the outputs is connected as one of the inputs to a gate which controls the other output. Unlike the combinational logic circuits considered previously, the circuit outputs depend on the inputs *and also on the outputs* – they are not just functions of the inputs.

Examination of Fig. 4.1a shows that the following relationships hold.

$$Q = \overline{S \cdot Q'} \quad \text{and} \quad Q' = \overline{R \cdot Q}$$

Application of de Morgan's theorem converts these into

$$Q = S + \overline{Q'} \quad \text{and} \quad Q' = R + \overline{Q}$$

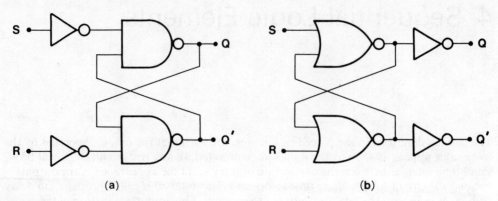

(a)   (b)

Fig. 4.1. The set–reset (SR) flip-flop

Similarly, the circuit of Fig. 4.1b obeys the equations

$$Q = \overline{\overline{S + \overline{Q'}}} \quad \text{and} \quad Q' = \overline{\overline{R + \overline{Q}}},$$

which can easily be reduced to the equations obtained for the circuit of Fig. 4.1a.

An analytical treatment of this type of circuit is not simple and instead of a detailed analysis the circuit is examined in all four possible cases.

(a) S = R = 0. This is the normal rest state of the circuit and the equations show that Q and Q' are different, i.e. $Q = \overline{Q'}$. There are two possible ways in which Q and Q' may differ; either Q = 1 with Q' = 0, or Q = 0 with Q' = 1. Both cases are stable ones; the term **stable** means that the circuit will remain in one state and will not change as long as the input conditions remain fixed, i.e. while S = R = 0. However, the circuit equations do not specify which state the circuit is in; this depends upon the previous history of the circuit. Thus, unlike a combinational logic circuit, the circuit equations of a sequential logic circuit do not always give a complete description of the outputs.

(b) S = 1, R = 0. The circuit equations give $Q = 1 + \overline{Q'} = 1$ and $Q' = 0 + \overline{Q} = 0$. Hence, these input conditions force Q to become 1. Any action which ensures that Q = 1 is a **SET** action, and this is why the S or SET input is so-called.

(c) S = 0, R = 1. This is the exact opposite of case (b) and the circuit outputs are Q = 0 and Q' = 1. An action causing Q to become 0 is a **RESET** action which is why the second input is the R or RESET input.

(d) S = R = 1. This is the most difficult case; the circuit equations give Q = Q' = 1 which results in a situation unlike the previous three cases where Q and Q' are always different. The situation is stable while S = R = 1 but a problem arises if S and R change simultaneously from 1 to 0. Examination of the circuit equations suggests that as the inputs become S = R = 0, the circuit should go to either of the possible $Q = \overline{Q'}$ states, but there is no information to indicate which one. A detailed examination of the operation of the individual gates in the circuit of Fig. 4.1a shows that if S and R go to 0 simultaneously while Q = Q' = 1, then both Q and Q' should become 0. However, if Q = Q' = 0 while S = R = 0 both Q and Q' should go to the 1 state and so on. Therefore, the circuit should become unstable and oscillate between the states Q = Q' = 1 and Q = Q' = 0.

In any real circuit, one of the output gates will operate marginally faster than the other and the circuit will not oscillate but will go to one of the $Q = \overline{Q'}$ states; which one of the two states cannot be predicted. In this particular case, the circuit action is indeterminate and a **race condition** or **hazard** is said to exist. Many different types of hazard may arise in logic circuits; the term is used whenever the output of a circuit cannot be predicted or when an incorrect output is produced (even if only for a brief time) before the circuit produces the correct one. A race condition is one particular type of hazard and arises when the final circuit output depends upon the relative operating speeds of two or more circuit components. Hazard conditions may occur in any sequential logic circuit; they must be identified and excluded in some way.

When circuits are constructed incorporating SR flip-flops the design should be such that the condition $S = R = 1$ never arises. In these circumstances, Q and Q' are always different and the outputs may be labelled Q and $\bar{Q}$.

4.2 The operation and use of an SR flip-flop

The use and action of an SR flip-flop can be described by the following statements.

(a) S and R are normally held at 0 and the outputs remain constant in either one of the $Q = \overline{Q'}$ states.
(b) An input sequence of 0 to 1 then back to 0 at the S input will ensure that $Q = 1$ and $Q' = 0$.
(c) A similar 0–1–0 input sequence at the R input ensures that $Q = 0$ and $Q' = 1$.
(d) In normal circuit design the input condition $S = R = 1$ should not be allowed.
(e) If power is connected to the circuit with $S = R = 0$, the circuit will take either one of the states $Q = \overline{Q'}$.

In some applications the initial state is important and an extra circuit is required to force the system into the required state. Such initialization problems are common in large systems; they are often overcome by incorporating a manual start switch into the system.

If these conditions hold, the circuit will always operate with the Q output as the inverse of Q'. The Q' output is usually labelled $\bar{Q}$; this is the most common nomenclature for commercial SR flip-flop elements.

The circuit diagrams of complicated sequential logic systems, particularly those constructed using transistor–transistor logic (TTL) elements, frequently include the circuit of Fig. 4.2. This is often referred to as an $\overline{SR}$ flip-flop; it has definite set and reset actions but all the inputs are the inverse of those in the description above. The inputs have been labelled $\bar{S}$ and $\bar{R}$ to distinguish them from those of the standard SR flip-flop. For this circuit the unallowed input condition is $\bar{S} = \bar{R} = 0$, and 1–0–1 transitions are required to

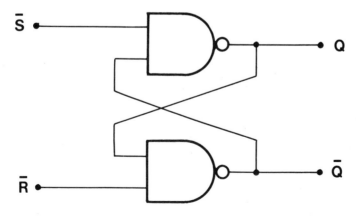

Fig. 4.2. The $\overline{SR}$ flip-flop

produce set and reset actions. This is a very useful circuit; one common application is as a switch buffer to overcome problems which arise when a mechanical switch is used with an electronic logic system (see Chapter 7).

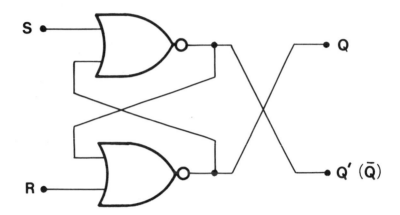

Fig. 4.3.

Another circuit is shown in Fig. 4.3. This one is constructed from NOR gates. The only difference between this circuit and those of Fig. 4.1 is that in the case S = R = 1 the circuit has the output Q = Q' = 0 instead of Q = Q' = 1. If the S = R = 1 input condition is not allowed, then the circuit behaviour cannot be distinguished from that of the circuits of Fig. 4.1.

4.3　The clocked SR flip-flop

Figure 4.4 is the circuit diagram of an SR flip-flop which has an additional input designated clock or enable. This input is used to change the SR flip-flop from an element

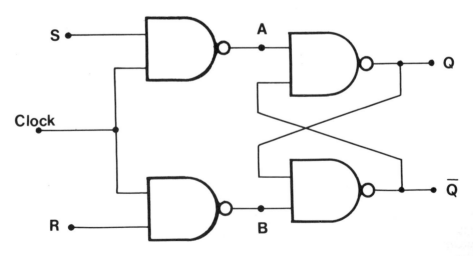

Fig. 4.4 Clocked SR flip-flop

used in asynchronous sequential circuits to one which may be used in synchronous circuits.

The clock input is the control input which may be used to cause the circuit to operate at a precisely determined time; this is the essential feature of a synchronous system. When the clock input is at logic 0 the points in the circuit diagram labelled A and B are both at 1 regardless of the inputs at S and R. The remainder of the circuit is a $\overline{S}\overline{R}$ flip-flop with A and B as the inputs. Therefore, as long as the clock input remains 0, the inputs to the $\overline{S}\overline{R}$ section stay at 1 and the final outputs remain constant at the present values of Q and $\overline{Q}$. Hence, while the clock is at 0 the S and R inputs have no effect on the circuit and may change many times (they may even take the $S = R = 1$ state) without affecting the outputs. Only when the clock input becomes 1 do S and R affect the output; when the clock is at 1, point A equals $\overline{S}$ and B equals $\overline{R}$, so that while the clock remains 1 the complete circuit behaves exactly as an SR flip-flop without a clock input.

To operate the flip-flop as a synchronous element, S and R are set to the required values while the clock input is 0. A 0 to 1 to 0 sequence (called a **clock pulse** or just a **pulse**) is then applied to the clock input with S and R kept constant. The flip-flop operates exactly as the non-clocked version except that any changes in the output states occur at a time determined precisely by the signal at the clock·input. If the same clock pulse is applied simultaneously to a number of flip-flops in a large system they will all operate at the same time. Circuits which have this common clock feature are more easily designed and have fewer operating problems than circuits without such a clock.

When a clocked flip-flop is used in this precisely timed manner its behaviour may be expressed in a tabular form. The form is similar to that of the truth tables used to describe combinational logic networks but differs in that the table for a clocked flip-flop shows the time dependence of the circuit. (This table is sometimes called a truth table, particularly in the data sheets supplied by integrated circuit manufacturers; a more suitable name for it is transition table or switching table.) The behaviour of a clocked SR flip-flop is given by Table 4.1.

Table 4.1

		t_n		t_{n+1}		
S	R	Q_n	$\overline{Q}_n$	Q_{n+1}	$\overline{Q}_{n+1}$	Comments
0	0	0	1	0	1	No change in outputs
0	0	1	0	1	0	
1	0	0	1	1	0	SET action
1	0	1	0	1	0	
0	1	0	1	0	1	RESET action
0	1	1	0	0	1	
1	1	0	1	?	?	Operation indeterminate
1	1	1	0	?	?	

In the table t_n denotes the time before a clock pulse and t_{n+1} denotes the time after the clock pulse; Q_n and Q_{n+1} are the values of Q which correspond to these times. An alternative notation which is sometimes used is Q_- for Q_n and Q_+ for Q_{n+1}. The comments have been added to the table to emphasize the circuit behaviour and, as indicated, the action in the case $S = R = 1$ cannot be predicted as it is the race condition again.

4.4   The D-type flip-flop (or latch)

If an inverter is connected so that the R input to a clocked SR flip-flop is always the
inverse of S then the complete circuit has a single input, D, and is known as a D-type flip-
flop. This is shown schematically in Fig. 4.5.

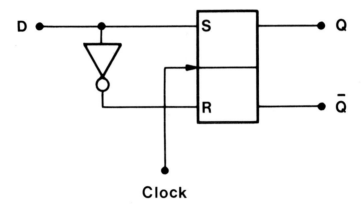

Clock

Fig. 4.5. The D-type flip-flop

Because R is always $\bar{S}$, the hazardous condition S = R = 1 cannot arise; the complete
action of the flip-flop is simpler than that of the SR flip-flop and is given in Table 4.2.

The D-type flip-flop will act as a storage element for a single binary digit (bit). The
logic states 1 and 0 may be used to represent the two binary digits 1 and 0 respectively. If
the bit to be stored is presented as the appropriate logic level at the D (data) input and a
clock pulse is applied to the flip-flop with the input maintained at D as long as the clock
input is at 1, then Q will become the same as D. Thereafter Q will remain at this value
until a new binary digit is input in the same way. In other words, a bit input at D is held
(stored) by the flip-flop even when it no longer exists at D.

Table 4.2

D	t_n		t_{n+1}	
	Q_n	$\bar{Q}_n$	Q_{n+1}	$\bar{Q}_{n+1}$
0	0	1	0	1
0	1	0	0	1
1	0	1	1	0
1	1	0	1	0

Some commercial D-type flip-flops do not behave in the manner described here. These
alternative flip-flops are of a type known as edge-triggered and should not be used by
inexperienced designers. Edge-triggered devices are introduced briefly later.

If a number of D-type flip-flops are connected to the same source of clock pulses then
the group of flip-flops will store several binary digits which may be used to represent a
multiple digit binary number. This group of digits is called a **word** and the group of flip-
flops is called a **register**. This grouping can be arranged in many ways and there are many
types of register.

4.5 The serial shift register

A simple shift register is shown in Fig. 4.6 and it consists of a number of clocked SR flip-flops connected in cascade (D-types could have been used). Three stages are shown but the system can be extended to as many as are required.

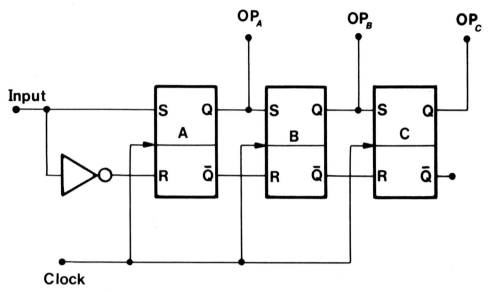

Fig. 4.6. A simple shift register

The circuit operation is quite simple; a new input is set up while the clock input is 0. A clock pulse then causes the new input to appear at the first stage output, OP_A; the original value of OP_A appears at OP_B and so on. Therefore, the complete contents of the register move along (shift) one flip-flop (stage) for each clock pulse applied. New data must be input to the first stage and the contents of the last stage are lost, although in some applications the output of the last stage is connected as the new first stage input and a recirculating action is produced.

There are no problems if each flip-flop operates slowly keeping its output unchanged

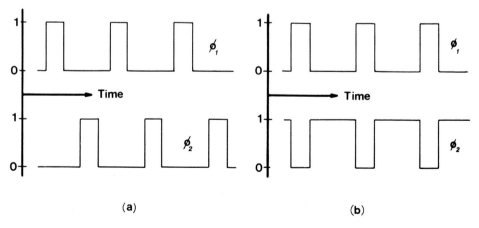

(a)   (b)

Fig. 4.7. Two phase clock sequences

until the internal components have changed. However, if the output of a flip-flop begins to change very soon after the clock pulse rises from 0 to 1, problems may arise. The output is also the next stage input, so the input to this following stage may change before its operation is complete (all stages are operating at the same time). If the input to a stage is uncertain in this way, then the stage and later ones in the shift register may not operate correctly. Modern electronic switching circuits operate very rapidly and this type of problem is common when they are used in this and similar applications. It is possible to deliberately slow down the system but such a solution is not satisfactory. In many applications high speed is required; furthermore when the complete register is a single integrated circuit delays are not an economical solution.

One of the best solutions to the problem is to double the number of flip-flops and to use two clock pulse sequences which have a constant relationship to one another in time. The two clock pulse sequences are known as a **two-phase clock** and two different sequences are shown in Fig. 4.7. Figure 4.7a shows a general purpose two-phase clock while Fig. 4.7b shows a more simply constructed form in which phase ϕ_2 is just the inverse of phase ϕ_1. The form shown in Fig. 4.7b may give problems in some applications and should only be used if the possible timing problems are understood and avoided.

A modified shift register using double flip-flops and a two-phase clock is shown in Fig. 4.8; again only three stages are illustrated.

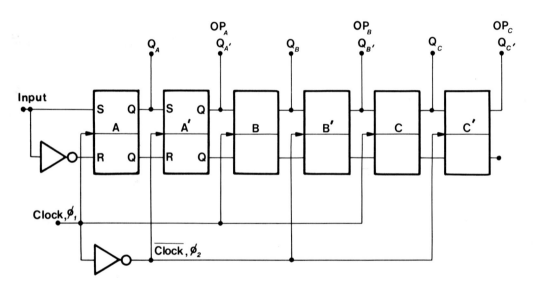

Fig. 4.8. A master-slave shift register

This arrangement is called a **master-slave** configuration (or sometimes a double-ranked one); the flip-flops A, B and C are the master flip-flops and A', B' and C' are the slaves. The rest condition is $\phi_1 = 0$ and hence $\phi_2 = 1$, so that the slaves are operating (enabled) but the masters are held inactive and their outputs remain constant. Since the input to A' is Q_A then the output of A' becomes the same as that of A, i.e. $Q_{A'} = Q_A$. Similarly, $Q_{B'} = Q_B$ and $Q_{C'} = Q_C$. When ϕ_1 becomes 1, then ϕ_2 becomes 0; the masters are now active and the slaves inactive. $Q_{A'}$, $Q_{B'}$, $Q_{C'}$ remain at the previous values but Q_A takes the value of the new input, Q_B becomes $Q_{A'}$, and Q_C becomes $Q_{B'}$. When ϕ_1 returns to 0, the masters are again inactive, and the slaves active, so that the new states at the Q outputs of the masters are transferred to the slave Q outputs which are also the external outputs. The two sets of flip-flops are clearly connected to operate in an alternating

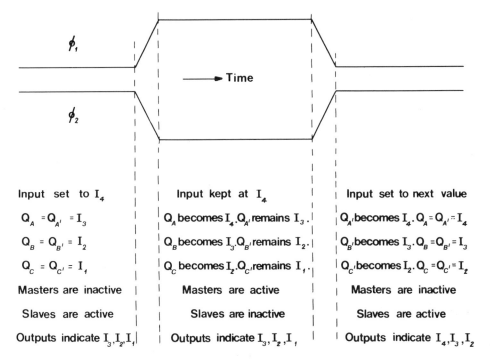

Fig. 4.9. Timing of a master-slave shift register

sequence in such a way that the inputs to any flip-flop remain constant while that flip-flop is active. This overcomes the problems associated with the previous circuit. The action is shown in combined graphical and tabular form in Fig. 4.9. I_4 represents the new input and I_3, I_2 and I_1 represent the initial values of $Q_{A'}$, $Q_{B'}$ and $Q_{C'}$ respectively.

4.6 Other registers

The shift register described above is one in which data is input one bit at a time to the first stage and the outputs of each stage are obtained at OP_A, OP_B, and OP_C. The outputs of all stages are available simultaneously – i.e. they may be connected to some other circuit in a parallel arrangement, but the inputs must arrive one after the other in a serial order in time along a single connection from the source. This type of register is a serial in-parallel out (SIPO) shift register.

Some integrated circuit shift registers are similar but the user can only obtain the output of the last stage (OP_C in the three-stage example) so that the output may only be obtained one bit at a time. This output is in serial form and the register is a serial in-serial out (SISO) shift register. Obviously the SIPO type may be used as a SISO type but the reverse is not possible.

In addition to being constructed with either serial or parallel ouput, registers may be constructed with serial or parallel input. A parallel input register is one which may have all the elements (stages) loaded simultaneously from separate (parallel) input connections instead of by repeated input of a single bit combined with a serial shift operation. Some parallel input registers are also capable of a shift action; the operation of such dual purpose registers depends on the status of a control input at the time at which a clock pulse is input to the register.

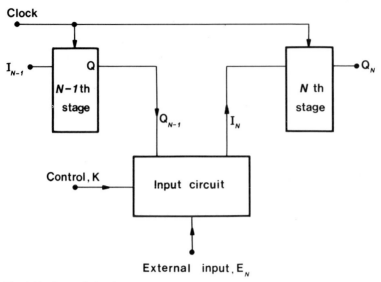

Fig. 4.10. Controlled register

The design of registers which have this controlled behaviour is simple; Fig. 4.10 is a block diagram for the Nth stage of such a register and shows a simple register with an additional section labelled 'input circuit'. The action of this input circuit is that if the control, K, is in one logic state then the register stage is loaded from the external input (parallel load) and if the control is in the other state the register shifts, i.e. the $N-1$th stage output is used as the Nth stage input. The control, $N-1$th stage output (Q_{N-1}), and the external input (E_N) form the inputs to the block called 'input circuit'. This block is a combinational logic circuit whose output is the Nth stage input, I_N. If K $= 0$ corresponds to the case of parallel loading and K $= 1$ corresponds to serial shifting, then the input circuit truth table, Table 4.3 can be constructed immediately.

Solution for I_N in terms of K, Q_{N-1} and E_N gives

$$I_N = K \cdot Q_{N-1} + \bar{K} \cdot E_N = \overline{\overline{K \cdot Q_{N-1}} \cdot \overline{K} \cdot E}$$

One input circuit is required for every stage (except the first) and the control, K, is connected as a common input to every input circuit.

Table 4.3

Control K	External input, E_N	$N-1$ stage output, Q_{N-1}	N stage input, I_N
0	0	0	0
0	0	1	0
0	1	0	1
0	1	1	1
1	0	0	0
1	0	1	1
1	1	0	0
1	1	1	1

Another feature which can be chosen when a shift register is designed is the shift operation itself. Some parallel in-parallel out (PIPO) registers have no shift operation at all, many registers shift in one direction only while others may be shifted in either direction. The design of registers which shift in either direction is similar to that of registers which can be parallel loaded or serial shifted; essentially the parallel input, E_N, is replaced by the output of the $N+1$th stage, Q_{N+1}.

When the number of stages of a register is specified, three other features must also be given to completely describe it. The input must be chosen to be serial or parallel (or both with a control); the output may be serial or parallel; there may be no shift operation, single direction shifting only, or shifting may be possible in either direction.

4.7   The JK flip-flop

This is the most complicated of the normally available flip-flops but it is probably the most important. One reason for its importance is that it may be connected to behave as any one of the other types introduced. Furthermore, the $J = K = 1$ condition (equivalent to $S = R = 1$ for an SR flip-flop) is not indeterminate but is defined to give a very useful change-over (toggle) action by which Q_{n+1} becomes $\bar{Q}_n$. The circuit has a clock input and two control inputs, J and K; the operation of the circuit is completely described by Table 4.4.

Table 4.4

J	K	t_n Q_n	$\bar{Q}_n$	t_{n+1} Q_{n+1}	$\bar{Q}_{n+1}$
0	0	0	1	0	1
0	0	1	0	1	0
1	0	0	1	1	0
1	0	1	0	1	0
0	1	0	1	0	1
0	1	1	0	0	1
1	1	0	1	1	0
1	1	1	0	0	1

Although Table 4.4 is a complete description of the behaviour of a JK flip-flop it is often useful to indicate the manner in which the bistable is operating. Table 4.5 describes the operation being performed by the bistable and it is convenient to refer to it as an **action table**.

The JK flip-flop is usually used in a master-slave form although alternative (non-master-slave) forms exist and are usually of a type called edge-triggered. Although the

Table 4.5

J	K	Action at next clock pulse
0	0	No change, $Q_{n+1} = Q_n$
1	0	SET, $Q_{n+1} = 1$
0	1	RESET, $Q_{n+1} = 0$
1	1	Change over (toggle), $Q_{n+1} = \bar{Q}_n$

term 'edge-triggered' has a precise meaning, it is often incorrectly used to describe any flip-flop which is not a master-slave type. Some edge-triggered devices have features which make them difficult to use and they should be avoided by the inexperienced designer. Some comments on master-slave and edge-triggered flip-flops are made in Section 4.8.

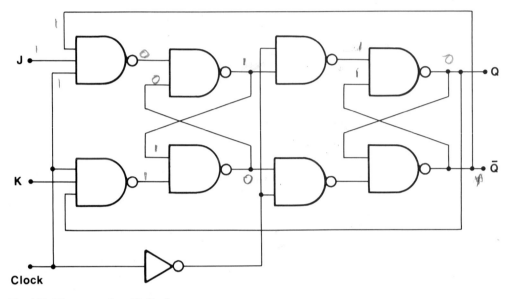

Fig. 4.11. The master-slave JK flip-flop

As in the case of the simple SR flip-flop, the JK type may be constructed in many ways. Figure 4.11 is the circuit diagram of a master-slave JK flip-flop which uses only NAND gates. The circuit is basically two clocked SR flip-flops connected in a master-slave configuration with the addition of further feedback. This additional feedback consists of cross-coupled connections from the slave outputs to extra inputs on the first gates of the master. There are thus two levels of feedback in this circuit rather than just one level. This makes analysis of the circuit behaviour difficult, but examination shows that there is no indeterminate condition and the useful toggle action is introduced.

If the circuit is examined in all possible circumstances, i.e. for all possible combinations of values of J, K and Q_n, then the behaviour given by Table 4.4 is obtained. This examination is left as an exercise for the reader who is interested in the detailed operation of this particular design of JK flip-flop.

Another way to express the behaviour of a clocked flip-flop is in a Boolean expression for the next output (i.e. Q_{n+1}) in terms of the inputs and the present output. In the case of the JK flip-flop, J, K and Q_n are regarded as input variables and Q_{n+1} is the output variable; a Boolean relation is then derived for Q_{n+1} in terms of J, K and Q_n. To produce a minimal form of Boolean expression a Karnaugh map may be constructed as shown in Fig. 4.12; the relationship obtained with the aid of the map is

$$Q_{n+1} = J \cdot \bar{Q}_n + \bar{K} \cdot Q_n.$$

A Boolean description can only be produced for those clocked flip-flops which have no indeterminate operations.

The exact time at which a clocked flip-flop operates, i.e. the time at which the new output appears, depends upon its construction and is different for master-slave and edge-triggered forms (see Section 4.8). In all the circuits which are considered in the following

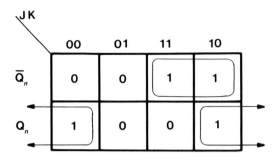

Fig. 4.12.

chapters only master-slave versions of clocked flip-flops are used; the Q and $\bar{Q}$ outputs only change when the clock input returns to 0. In other words, the new output only appears after the full 0–1–0 clock pulse sequence; until the pulse is complete, the outputs remain at their previous values. If J and K change while the clock is in the 1 state, the final circuit action will depend upon the internal construction of the flip-flop which is usually unknown (e.g. the exact internal form of integrated circuit flip-flops is rarely supplied by manufacturers). Unless detailed descriptions of the circuit behaviour are available, J and K should be set to the required values while the clock is at 0, and should remain constant throughout the clock pulse.

Many integrated circuit JK flip-flops have multiple J and K inputs. In most cases, J is defined to be the AND function of all the J inputs. Thus if the inputs are J_1, J_2 and J_3, then the value of J which determines the flip-flop operation is given by $J = J_1 . J_2 . J_3$. Similarly, K is usually the AND function of all the K inputs. However, it is important to note that *this is not always the case* and it is necessary to read the specification of multiple-input devices very carefully.

Another common feature of integrated circuit flip-flops is the provision of CLEAR and PRESET inputs. There is no general convention which specifies the logic level required to activate such inputs; some devices require a 1 while others require a 0. However, the action is such that when the appropriate level is applied to the CLEAR input, the Q output immediately becomes 0, regardless of all other inputs. An input to PRESET results in $\bar{Q}$ being forced immediately to the 0 state. In general, the condition that the Q and $\bar{Q}$ outputs are different is retained, but if both CLEAR and PRESET are activated simultaneously a problem arises; there is no standard circuit behaviour, but in may cases both Q and $\bar{Q}$ are forced to 1 while both inputs are active.

The purpose of the CLEAR and PRESET inputs is not to provide a flip-flop with more modes of operation in normal circuit constructions. They should only be used to initialize circuit conditions; i.e. they are connected so that the user of a complex circuit may force it into a known starting condition. Any other use of the CLEAR and PRESET inputs requires knowledge of the many problems which arise because these control inputs have an immediate effect whereas the other control inputs only have an effect at the time at which the clock input changes.

4.8   Comments on master-slave and edge-triggered flip-flops

Many readers will find that this section should be omitted on a first reading as comparisons of these two types of flip-flop are often found to be confusing. Some of the difficulties arise because there are variations in the way in which different designs of edge-triggered devices behave. Also, the term is sometimes used to describe devices which are not true edge-triggered ones. For example, the forms of single (non-master-slave) SR and

D-type flip-flops introduced previously are sometimes called edge-triggered; they are not edge-triggered designs but superficially appear to behave as if they are.

True edge-triggered devices are such that the values of the control inputs (i.e. J and K for a JK flip-flop) at the time of the clock transition affect the output immediately according to the state table for the device. Values of the control inputs at all other times have no effect on the flip-flop. If output changes occur on the 0 to 1 clock transition, the flip-flop is positive edge-triggered and if changes take place at the time of the 1 to 0 clock transition, it is negative edge-triggered. The positive version is the more common one.

In circuits which incorporate several bistables edge-triggered types are often difficult to use. One reason is that the outputs change as soon as the clock transition occurs and sometimes they cannot be used to control other flip-flops connected to the same clock pulse. This problem is similar to the one described in Section 4.5 when shift registers were introduced. With the exception of using simple D-type flip-flops for latching circuits, for example PIPO none shifting registers, the inexperienced designer should always use master-slave flip-flops when clocked devices are required.

The master-slave form (sometimes called pulse-triggered) has the advantage of a double action. The conditions at the control inputs affect the master while the clock is high, but the output of the master is only transferred to the slave (and hence the external output) when the clock goes low. Therefore, a complete clock pulse, not just one edge, is required to operate a master-slave flip-flop. In addition to the design problems encountered with edge-triggered devices, the other obvious differences between them and master-slave devices is that the output appears after the falling edge of the clock pulse for master-slave flip-flops, whereas most edge-triggered devices are positive ones and the output appears after the rising edge.

The range of trigger modes available requires that the user of integrated circuit flip-flops must read and understand the data supplied by the manufacturer. This may be difficult as the manufacturer's data is sometimes incomplete or ambiguous. It is reasonable to assume that – with only a few exceptions – all master-slave flip-flops will behave in the manner described here, but it is essential to have a full knowledge of the specification of edge-triggered devices. The edge-triggered form operates as a single flip-flop whereas the master-slave form has two in series. Therefore, an edge-triggered version is usually twice as fast as a master-slave form.

Because of timing problems, even experienced designers tend to use master-slave devices unless the application requires the higher speed of edge-triggered versions. The two types should not be used within a single system unless all the timing problems which can arise are fully understood.

4.9   Problems

1   Use a method similar to that used in Section 4.1 to determine the complete behaviour of the NOR gate SR flip-flop shown in Fig. 4.3.

2   Design the Nth stage of a shift register which can be shifted in either direction. The direction of a shift is determined by the state of a control input.

3   Design the Nth stage of a universal register; i.e. a register which can be shifted in either direction or loaded in parallel. (Hint: – two control inputs are required.)

4   Analyse the circuit of Fig. 4.11 by choosing one possible condition at some instant and determining the logic levels throughout the circuit before, during, and after the clock pulse. Repeat for all possible conditions and hence show that the circuit does behave as a JK flip-flop.

5   Figure 4.13a represents a four-stage serial shift register which is constructed using SR master-slave flip-flops. The Q and Q̄ outputs of the last stage are cross-connected to the first-stage inputs as illustrated. If all the stages are initially zero (i.e. Q outputs zero)

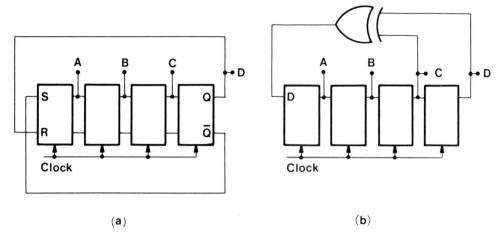

Fig. 4.13.

determine the register contents after each clock pulse input for at least ten clock pulses.

Tabulate your results and comment on the circuit behaviour.

6　Figure 4.13b represents a four-stage serial shift register which is constructed using D-type master-slave flip-flops. The Q outputs of the last stage and the preceding stage are inputs to an exclusive-OR gate and the output of this gate is the first-stage input. Initially, the first stage contains a 1 (i.e. Q is 1) and all other stages are zero; determine the register contents after each clock pulse input for at least sixteen clock pulses.

Tabulate your results. What happens if the initial condition is that all stages *including* the first are zero?

5 Sequential Logic Systems

In Chapter 4, sequential logic circuits were classified as either synchronous or asynchronous. Large systems generally consist of a mixture of both types of circuit and should be considered as an assembly of sub-systems, each of which is of one type. These sub-systems are connected together so that the overall behaviour appears to be completely synchronous or completely asynchronous.

For example, most computers have a master clock and are constructed so that all operations take place at times which are precisely determined relative to this clock. Such computers appear to be synchronous systems. Within a computer there are many sub-systems; one of these is an arithmetic unit which performs the calculations and is usually synchronous. Another sub-system is the device by which information (programs, data, etc.) is input to the computer; this is inherently asynchronous because the time at which a user initiates some operation is entirely random with respect to the internal master clock of the computer.

Synchronous circuits are usually easier to design than asynchronous ones, but in the case of pure binary counters asynchronous design techniques are particularly simple. These are the only asynchronous circuits which will be considered in detail.

5.1 Counters

The term **counter** has a special meaning when used to describe a sequential logic circuit. It refers to a circuit which has a single input, several outputs and behaves in the following manner. Every time a pulse (clock pulse) is input to a counter circuit, one or more of the outputs changes. The circuit follows some sequence in which no combination of outputs is repeated until the first one recurs after some fixed number, N, of input pulses – hence there are N different output combinations. When the circuit returns to the original combination of outputs further input pulses cause it to follow exactly the same sequence.

This circuit is called a 'divide-by-N counter', because if it is arranged (with additional circuits) to produce a single output pulse for every complete cycle of N input pulses, then the number of output pulses is the number of input pulses divided by N. Although these circuits are called counters, their use is not restricted to counting applications; they form the basis of many sequential logic circuits. For example, each of the different combinations in the N-step sequence may be arranged to initiate some operation in a piece of equipment; N different operations are then performed in a set order. This sequence is called a program and this type of program is typical of those used in simple automatic equipment such as traffic light controllers, automatic clothes washing machines, etc.

Although counters require only one input, many commercial counters have additional inputs called reset (or zero) and preset. The reset input is used to force set the counter to the first position in the sequence regardless of its existing position or input; this is most frequently the condition with all the circuit outputs zero. A preset input sets the counter to some other point in the sequence; usually it is the last one before it returns to the initial zero position.

A **pure binary counter** is one in which the outputs may be regarded as a multiple digit binary number; the initial output is the one with all the binary digits (circuit outputs) zero and the counter sequence is such that the binary number increases by 1 for each input pulse until all the outputs are 1 (i.e. after $2^n - 1$ input pulses when the counter has n outputs). The next input pulse returns the counter to zero. Hence, 2^n input pulses are required for a complete sequence of an n output (n-stage) pure binary counter. The behaviour of a pure binary counter with three outputs, A, B and C, is given in Table 5.1. Output A represents the least significant binary digit and C represents the most significant digit; as $n = 3$ the counter is a divide-by-eight one.

Table 5.1

C (2^2)	Outputs B (2^1)	A (2^0)	Number of input pulses received.
0	0	0	0 (start)
0	0	1	1
0	1	0	2
0	1	1	3
1	0	0	4
1	0	1	5
1	1	0	6
1	1	1	7
0	0	0	8

Binary counters which are *not* classified as pure binary ones are similar to the pure binary type; they are divide-by-N counters which start at zero and follow a binary counting order, but they return to zero after N input pulses, where $N \neq 2^m$ (m is any integer).

Non-binary counters do *not* follow a binary counting sequence; they follow some other fixed sequence which obeys the rule that no combination of outputs occurs more than once in a cycle of N input pulses and the counter returns to its initial (zero) state after N input pulses. Table 5.2 is an example of one possible divide-by-six non-binary counter; it is one which changes in a Gray code sequence.

The binary counters described are such that the number represented by the outputs increases with each input pulse; consequently these are known as **'up' counters**. Alternative designs are possible in which the binary number decreases by one for each

Table 5.2

C	Outputs B	A	Number of input pulses received.
0	1	0	0 (start)
1	1	0	1
1	0	0	2
1	0	1	3
0	0	1	4
0	1	1	5
0	1	0	6

input pulse and these are called **'down' counters**. Both types may be combined into a single counter called an **'up–down' counter**; the direction of the count is usually controlled by the logic state at an extra control input. Non-binary counters may also be specified as up or down types if the outputs correspond to the sequence of a well-defined code.

5.2   Pulse sequences

Sequential circuits require some form of pulse input, and although the form of sequence supplied may be complicated, it will usually be one of two types. The sequence may be one in which the pulses occur at intervals which are entirely random in time as in Fig. 5.1a; the pulse width in such a sequence may be random or constant and is usually unimportant. Alternatively, the sequence is one which is constant in time and repeats at a fixed interval; it may be a single pulse as in Fig. 5.1b or a more complex scheme as in Fig. 5.1c.

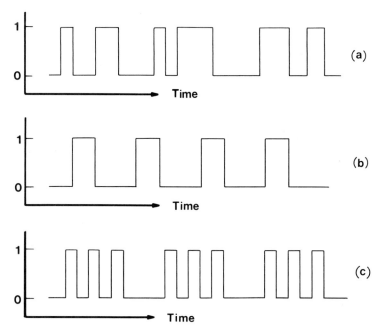

Fig. 5.1. Pulse sequences

The sequence in Fig. 5.1a is similar to the type generated by a detector of random events; e.g. people passing through an entrance, vehicles travelling along a road, or nuclear decay of a radioactive material. If a detector is arranged to indicate such events, then a counter circuit connected to the detector output may be used to determine the number of events.

A regular single pulse sequence of the form shown by Fig. 5.1b is usually called a **clock pulse sequence** or clock pulse train. It is frequently used as the input to a counter whose outputs control a sequence of events which occur at precisely determined intervals in time.

5.3   Asynchronous pure binary counters

If the J and K inputs of a JK flip-flop are set to 1, then it behaves as a divide-by-two counter. Because $J = K = 1$, each clock pulse input to the flip-flop causes the Q output to change. To produce a 0–1–0 sequence at Q with J = K = 1 requires Q to change twice; i.e. two input clock pulses produce an output sequence of 0–1–0 which is itself a pulse. Thus, one output pulse is generated for every two input pulses and the circuit meets the definition of a divide-by-two counter.

If several–say n–JK flip-flops are connected in cascade by using the Q output of one as the clock input to the next, then a divide-by-N pure binary counter is produced for which $N = 2^n$. A three-stage counter constructed in this way using master-slave flip-flops is shown in Fig. 5.2, which also includes the **timing diagram** for an ideal circuit. Timing diagrams are a useful aid when describing sequential logic circuits and show the logic states at different points in the circuit on a common time scale.

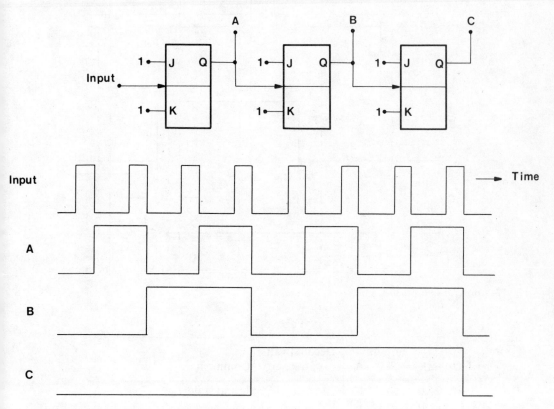

Fig. 5.2. Asynchronous divide-by-eight counter

Figure 5.2 indicates that a single output pulse (0–1–0 sequence) is produced at C when eight pulses are input to the counter circuit. The circuit operation is very simple; every input pulse causes A to change and until A changes there is no input to B, so the latter cannot operate. B changes every time that A goes from 1 to 0 (because master-slave flip-flops are used) and whenever the change in B is from 1 to 0 this causes C to change. Thus, the action propagates through the counter. Each stage must wait for a change in the previous one before it can begin to change and the circuit is clearly asynchronous. This particular design is often called a ripple-through counter or a ripple counter. The order in which the outputs of a counter change is shown by a state table; that for a three-stage ripple counter is the same as Table 5.1.

Ripple counters are adequate for many applications and many integrated circuit counters are of this type, but one feature of these counters may cause problems in some applications. The problems arise because each flip-flop takes a finite time to operate (i.e. there is a propagation delay) and the timing diagram in Fig. 5.2 is not strictly correct. If all the flip-flops are assumed to have the same propagation delay, t_d, and if the output is assumed to change instantaneously after this delay, then a corrected timing diagram may be drawn. Figure 5.3 shows part of the corrected diagram.

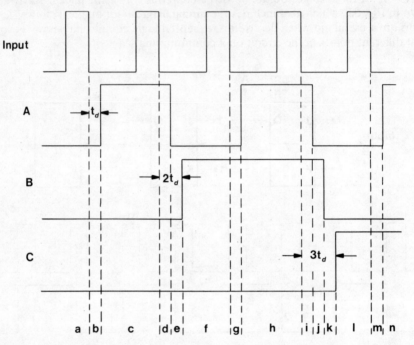

Fig. 5.3. Corrected asynchronous counter timing diagram

Table 5.3

Time	Outputs			Decimal value of output	Comments
	C	B	A		
a	0	0	0	0	Initial state
b	0	0	0	0	
c	0	0	1	1	Correct state but late by t_d
d	0	0	1	1	
e	0	0	0	0	Transient for interval t_d
f	0	1	0	2	Correct state but late by $2t_d$
g	0	1	0	2	
h	0	1	1	3	Correct state but late by t_d
i	0	1	1	3	
j	0	1	0	2	Transient for interval t_d
k	0	0	0	0	Transient for interval t_d
l	1	0	0	4	Correct state but late by $3t_d$
m	1	0	0	4	
n	1	0	1	5	Correct state but late by t_d
		etc.			

With the aid of Fig. 5.3, a table can be constructed which shows the exact circuit behaviour; Table 5.3 is part of the exact table. Examination of this table shows two features of the circuit which were not present when propagation delay was ignored. Firstly, when a correct output state is generated, it may occur later than it should by a time of up to $n \times t_d$, where n is the number of flip-flops in the counter. Secondly, some output states arise which are not in the correct sequence; they occur only for a brief period which is usually a single propagation delay. Incorrect outputs of short duration are called **transients** or transient outputs and are an additional type of **hazard** situation. Unlike the hazards described previously, these do not cause indeterminate or incorrect circuit operation; the circuit eventually reaches the correct state, but the incorrect transient outputs may cause problems in large systems.

5.4 An application of an asynchronous counter

Pure binary and other binary asynchronous counters are frequently used as components in large sequential systems. In electronic systems the transients occur for a very short time and in many applications they do not cause problems. Figure 5.4 shows a typical circuit which incorporates a divide-by-four asynchronous counter; this particular circuit is a controller for a scanner.

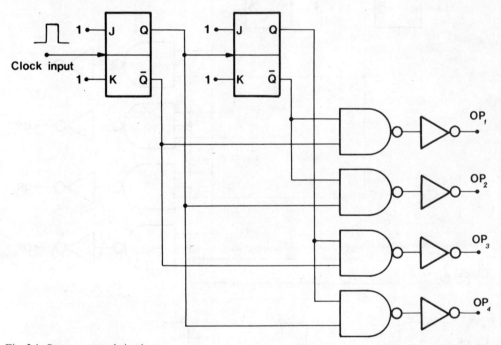

Fig. 5.4. Scanner control circuit

The circuit has four outputs, each of which corresponds to one state of the counter. If there is no propagation delay, each output goes to 1 in turn and no two outputs are ever 1 at the same time. A typical use for this circuit is in a monitoring system when readings of four quantities are taken using a single recording instrument which is switched to each signal source in turn. Such a system may be used, for example, to record temperatures at different positions in a piece of industrial equipment with a single chart recorder, alarm system or other device. Four temperature sensors (e.g. thermocouples) are switched in turn to the recording device; the switches are controlled by the outputs of the scanner circuit whose clock input is a pulse generator running at a constant frequency which determines the interval between readings.

If the switches used are relays and the scanner driving them is an electronic circuit, the transients are unimportant because they will last for less than 0.1 microseconds, whereas the fastest relays take about 1 millisecond to operate. However, if the switches are themselves electronic devices, it is probable that they will operate in a time which is shorter than the duration of a transient output, so the circuit may operate incorrectly.

5.5   Elimination of transient outputs

The most common solution used to remove transient outputs produced in asynchronous circuits is to use a multi-phase clock. For a circuit which is as simple as the scanner in

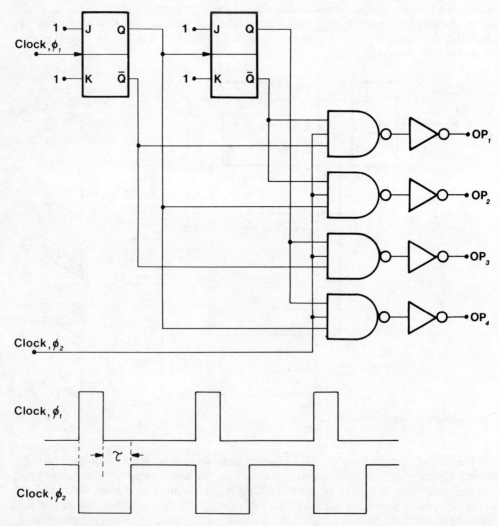

Fig. 5.5. Transient suppressed scanner control circuit

Fig. 5.4, a two-phase clock is adequate. As the scanner is designed for applications where outputs of 0 do nothing, then if all outputs are simultaneously zero there are no problems and clock phase ϕ_2 is used to switch off all the outputs until the circuit has settled into its correct output state. Figure 5.5 shows the circuit for the modified scanner and the timing diagram for the two-phase clock. The main disadvantage of this construction is that the delay, τ, between the two clock pulses must be at least the maximum period for which transients can exist, so this type of circuit is inherently slow in operation.

Switching the outputs from an asynchronous circuit so that they appear at a precisely determined time gives the circuit an apparent synchronous behaviour. This technique forms the basis of many methods which are used to connect asynchronous circuits into synchronous systems. Note that there is a small propagation delay in the output NAND gates and inverters. Such delays occur in synchronous circuits and even well-designed synchronous circuits exhibit a slight asynchronous behaviour.

In those cases where the outputs cannot all be switched off until propagation delay effects have disappeared, a two-phase clock solution may still be used. The circuit is built as in Fig. 5.4, but the outputs are used as the inputs to a parallel in-parallel out register, and the final outputs are obtained from the register outputs. Phase ϕ_2 is now used as the clock input to the register and is used to load the register when the circuit has settled in the correct output state.

6 The Design of Sequential Logic Systems

Many methods exist for the systematic design of sequential logic circuits; one is described here and is based on a simplified form of Moore's model of sequential circuits. The technique introduced produces a synchronous circuit which is reliable, although it is not always the most economical one possible for a particular application. However, since the design method produces circuits which meet the specification and exhibit very few operating problems it is a useful method for both inexperienced and experienced designers.

6.1 Some definitions

In order to describe any sequential circuit design technique it is useful to define some features of sequential logic circuits in a precise manner.

(a) States or internal states. A sequential circuit must include one or more internal memory elements; these are bistables (flip-flops) and are denoted by $A_1, A_2, \ldots, A_i$ where i is the number of bistables in the circuit. At any instant in time the circuit may be defined to be in some state (or internal state) denoted by S_x. Each state corresponds to a different set of the Boolean values (logic levels) of the Q outputs of all the bistables. Therefore, the system may take any one of the states $S_1, S_2, \ldots, S_j$, where each S_x corresponds to one unique set of values of the outputs of $A_1, A_2, \ldots, A_i$. In general, if a system includes i flip-flops, it will have 2^i possible internal states and these can be shown in a **state assignment table**. Table 6.1 is one of the ways in which the states could be allocated for a system with two bistables; since $i = 2$, in this case there are four states, S_1, S_2, S_3 and S_4, which may be allocated in any way at all. It is often convenient to allocate the states in order, with the bistable outputs arranged to represent the digits of an increasing binary number as in Table 6.1. However, allocation in this way is not essential.

Table 6.1

| Q outputs | | State |
A_2	A_1	allocated
0	0	S_1
0	1	S_2
1	0	S_3
1	1	S_4

In many applications, the number of states required is not exactly 2^m (where m is integral). For example, a divide-by-five counter requires five states. This is not important; circuits are designed with enough bistables to produce at least the required number of states, and initially any extra states are neglected. If a system requires j states then the circuit must include at least n flip-flops where $2^n \geqslant j$. In the design method which follows, the smallest number of flip-flops possible is used. Therefore, n is as small as possible, which implies that $2^n \geqslant j > 2^{n-1}$ and hence when j is known n can be determined

Note that a bistable itself is a sequential logic circuit which has two internal states; the definition here of the state of any sequential circuit is consistent with the description of the state of a bistable made in Chapter 4.

(b) Input conditions. A synchronous sequential logic circuit must always have a clock input; in addition, it may have a number of **control inputs**, $B_1, B_2, \ldots, B_k$, although many circuits have no control inputs. The clocked bistables described in Chapter 4 are synchronous circuits with control inputs; for example, the J and K inputs of a JK flip-flop determine (control) its action when the next clock pulse is input to it.

The logic levels at the control inputs may be used to define **input conditions**, $I_1, I_2, \ldots, I_l$, where each I_β corresponds to a unique set of Boolean values of the control inputs. Therefore, if there are k control inputs there will be 2^k different input conditions; unlike internal states whose number need not be exactly 2^n, there are always 2^k input conditions.

(c) Circuit outputs. In most applications, the outputs of a sequential logic circuit are used to control the condition of some external system. This control of conditions elsewhere may be achieved directly by the Q outputs of the flip-flops in the sequential circuit, but in many cases additional circuits are needed to produce the outputs required. These output circuits have the flip-flop Q outputs as their inputs and are combinational logic networks for which design techniques have been described previously. The scanner circuit connected to the divide-by-four asynchornous counter described in Section 5.3 is one such output circuit.

Except to note that combinational logic circuits are often required to provide the final outputs of a sequential logic circuit, these output circuits will be ignored at present. Such output circuits may be designed when the design of the sequential section of the network has been completed.

6.2 State diagrams

As in all design problems, a complete specification of a sequential logic circuit is required before any attempt can be made to produce a circuit design. A **state diagram** is a useful aid when describing any sequential logic circuit as it is a pictorial representation of the circuit operation.

To construct a state diagram, it is necessary to know the number of states and the number of input conditions. Each state is allocated a unique symbol and corresponds to one node in the diagram; it is represented by a circle containing the symbol for that state. From each node there must be a separate flow line for every possible input condition. The flow line has an arrow which shows its direction and is labelled with a symbol to indicate the input condition it represents. The line ends at the node (state) to which the circuit will change when the next clock pulse is input to the circuit. It is essential to show flow lines from every node for every possible input condition, although if two lines are identical, i.e. they are between the same two nodes *and* are in the same direction, they may be combined *but are labelled with both input conditions*. A flow line must be shown even in those cases for which the circuit does not change state, or cases which it is considered will never arise; these flow lines leave the node and loop back to it.

To determine the behaviour of some system from its state diagram when a clock pulse is input, all that is necessary is to identify its present state and to follow the flow line for the input conditions which exist. This line will end at the node which corresponds to the state to which the system changes when the next clock pulse input is applied to the circuit.

Example 6.1

Construct state diagrams for two circuits both having three states; S_1, S_2 and S_3; and a single control input which gives rise to two input conditions of I_1 and I_2.

(a) The first circuit is a divide-by-three up counter which behaves normally when the control input is 0 and it resets to state S_1 on the next clock pulse if the control input is 1.
(b) The other circuit is an up–down divide-by-three counter whose direction is determined by the logic value of the control input.

Solution

In case (a) the input conditions must be assigned. Arbitrarily choose I_1 to be the input condition when the control input is 0, and I_2 to be that when the control input is 1.

In case (b), the choice of control inputs to cause upward and downward counting is left to the designer. Choose a control input of 0 to correspond to upward counting and let this be input condition I_1; the other possible control input of 1 must correspond to downward counting and may be called input condition I_2.

In both cases, choose upward counting to be the sequence S_1 to S_2, S_2 to S_3 and S_3 to S_1 (this is just the assignment of states). It is now a simple exercise to draw the two state diagrams; in each case the three symbols for the states are drawn and then flow lines corresponding to I_1 and I_2 are drawn from each state symbol. The completed diagram for case (a) is Fig. 6.1a and that for case (b) is Fig. 6.1b.

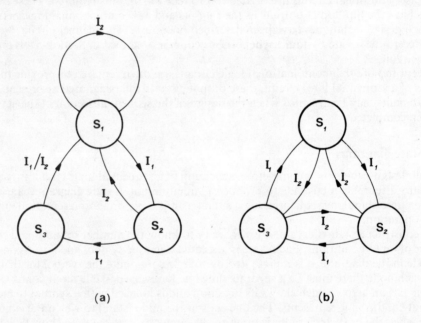

(a)   (b)

Fig. 6.1.

Example 6.2

Devise a state diagram for a JK flip-flop. (Since it was stated in Section 6.1 that any bistable is just one particular sequential logic circuit, it must be possible to draw a state diagram for any clocked bistable.)

Solution

The flip-flop has two states of $Q = 0$ and $Q = 1$ and has J and K control inputs which will generate four input conditions. The allocation of input conditions and action of the flip-flop for each one are shown in the tables included in Fig. 6.2. The state diagram may be drawn immediately from the information in the table and is also shown in Fig. 6.2. Note how the 'no change' cases are indicated by looped flow lines which leave a state and return to the same state.

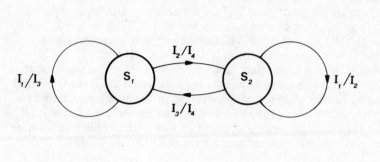

State Allocation

State	Q
S_1	0
S_2	1

Input Conditions

J	K	Condition
0	0	I_1
1	0	I_2
0	1	I_3
1	1	I_4

Fig. 6.2. State diagram for a JK flip-flop

6.3 State tables

The state diagram is a pictorial device which provides an exact and easily interpreted description of the behaviour of a sequential logic circuit. However, it is usually difficult to design a circuit directly from a state diagram; a **state table** (transition table) is a more useful aid and may be developed directly from the state diagram for a circuit. State tables may also be constructed without first drawing state diagrams, but there is a much greater possibility of errors being made when there is no state diagram.

In its simplest form, a state table consists of a column for the present state, a second column for the input condition, and a third column which shows the state to which the circuit will change when the next clock pulse is received by the circuit. Each row in the state table corresponds to one combination of initial state and input condition. *There must be a row for every possible combination of initial state and input condition*, so that for a circuit with j states and l input conditions there must be $j \times l$ rows in the table. The state diagram of a divide-by-three counter with reset was developed in Example 6.1a; Table 6.2 is the state table for this counter, and has six rows because the counter has three states and two input conditions.

Table 6.2

Present state	Input condition	Next state
S_1	I_1	S_2
S_1	I_2	S_1
S_2	I_1	S_3
S_2	I_2	S_1
S_3	I_1	S_1
S_3	I_2	S_1

This form of the state table is just a tabular version of the state diagram. A more useful version is one which indicates logic levels. In other words, each S_α is replaced by the values of the Q outputs of all of the bistables in the circuit, and each I_β is replaced by the logic levels at all the control inputs. In order to construct this form of state table, it is necessary to relate the logic levels at the Q outputs to the states, and those at the control inputs to the input conditions (if these assignments have not already been made). For example, the divide-by-three counter with reset will incorporate two flip-flops; suppose that their Q outputs are A and B, then choose S_1 to be $A = B = 0$, S_2 to be $A = 1$, $B = 0$, and S_3 to be $A = 0$, $B = 1$. If the control input is Z, then the original specification in Example 6.1 required that $Z = 0$ is input condition I_1 and $Z = 1$ is I_2.

To produce the extended version of the state table, the single column for the present state is replaced by a group of columns which indicate the logic levels at all the Q outputs (one column per flip-flop). The next state column is replaced in a similar manner, and the input condition column is replaced by a group of columns showing the logic levels at all the control inputs. Using the state and input allocations chosen for Example 6.1a, its state table, Table 6.2, may be converted into this extended form and becomes Table 6.3.

Table 6.3

Present state		Control input(s)	Next state	
B	A	Z	B	A
0	0	0	0	1
0	0	1	0	0
0	1	0	1	0
0	1	1	0	0
1	0	0	0	0
1	0	1	0	0

This table completely specifies the required circuit behaviour. Although the table appears to be a truth table (and is sometimes called one) it is not one because in a single row of a state table some entries correspond to present logic values of circuit outputs and other entries correspond to future values. A single row in a truth table should only show logic values which exist at the same instant in time.

State diagrams and state tables allow sequential logic circuits to be completely and accurately described; the next step is to introduce a method by which these can be used to produce a design for the circuit.

6.4 Development of an excitation table

Only JK master-slave flip-flops may be used as the bistable elements in the design method described here. As the circuit is to be a true synchronous one *all* of the flip-flops must have their clock inputs connected directly to a single common source of clock pulses, so that all the flip-flops which are to change on a particular clock pulse do so simultaneously. State tables such as Table 6.3 show how the Q outputs of all the bistables must change when the next clock pulse is input to a circuit which is in a specified state with known input conditions.

For each bistable Q output four different situations may be indicated in the state table. If Q is 0, then it may be required to remain 0, or it may be required to change to 1 when the next clock pulse is input. Similarly, if Q is 1 it may remain 1 or change to 0. All four possible cases will be called **transitions**, even though two of the cases involved no change in the Q output. The action of a JK flip-flop for different values of the J and K inputs was summarized in Table 4.5 which is reproduced here for reference.

When the present value of Q is 0 and the state table requires that it remains zero when a clock pulse is input, there are two ways in which this can be arranged. If

Table 4.5

J	K	Action at next clock pulse
0	0	No change, $Q_{n+1} = Q_n$
1	0	SET, $Q_{n+1} = 1$
0	1	RESET, $Q_{n+1} = 0$
1	1	Change over (toggle), $Q_{n+1} = \bar{Q}_n$

$J = K = 0$ then no change will take place and Q will remain at 0. Alternatively, if $J = 0$ and $K = 1$, a reset will occur and ensures that Q becomes 0 (i.e. remains 0). Hence, if Q is 0 and a value of $Q = 0$ is required after the next clock pulse, it is necessary to have $J = 0$, but K may be either 0 or 1. This is another type of 'don't care' situation; K may be 0 or 1 – a 'don't care' – not because the conditions will never arise, but because with either value at the K input the bistable will operate as required by the state table.

Thus, to obtain the output transition of 'Q is 0 and becomes 0', the control inputs required by a JK flip-flop are $J = 0$ and $K = $ 'don't care' $= X$. A similar argument for the transition 'Q is 0 and becomes 1' gives the requirement that the control inputs are $J = 1$ and $K = X$. Examination of all four possible output transitions to determine the values of J and K required to produce each one leads to the results summarized in Table 6.4.

Table 6.4

Present Output Q_n	Next Output Q_{n+1}	Control Required J	K
0	0	0	X
0	1	1	X
1	0	X	1
1	1	X	0

Table 6.4 is the **excitation table** for a JK flip-flop and contains all the information required to extend the state table of a circuit so that it becomes an excitation or switching table. An excitation table for a sequential logic circuit constructed using JK flip-flops shows the logic levels required at every J and K input to produce the correct Q outputs when the next clock pulse is input. If J_A and K_A are the control inputs of the flip-flop whose Q output is called A in Table 6.3, then using Table 6.4, the values of J_A and K_A needed to produce the required change in A may be determined for each row in the truth table. For example, in the first row of Table 6.3 output A changes from 0 to 1; examination of Table 6.4 indicates that to produce this transition, J_A must be 1 and K_A is a 'don't care'. In the same row of the table, B is required to make the 0 to 0 transition which requires $J_B = 0$ and $K_B = $ 'don't care' $= X$.

By using this technique of examining how each flip-flop Q output is required to change in every row in a state table, the corresponding excitation table can be produced. This is a state table to which further columns have been added; these columns show the values required at all the J and K inputs to produce the specified changes in the Q outputs. In the case of Example 6.1a, for which Table 6.3 is the state table, this determination of J and K inputs produces Table 6.5 as the excitation table.

One step remains to complete this table; the example requires two bistables. Hence,

Table 6.5

Present State B A	Control Z	Next State B A	Flip-flop controls J_B K_B J_A K_A			
0 0	0	0 1	0	X	1	X
0 0	1	0 0	0	X	0	X
0 1	0	1 0	1	X	X	1
0 1	1	0 0	0	X	X	1
1 0	0	0 0	X	1	0	X
1 0	1	0 0	X	1	0	X

Table 6.6

Present State B A	Control Z	Next State B A	Flip-flop controls J_B K_B J_A K_A			
1 1	0	X X	X	X	X	X
1 1	1	X X	X	X	X	X

four states are possible, but only three are used. To complete the circuit design, it is necessary to specify values for all the J and K inputs corresponding to all possible combinations of initial state and input condition, including the impossible condition of the circuit being in the unused or redundant state. (Other systems may have several unused states or none.) At present, it is assumed that the circuit can never be in this unused state, so that it does not matter what values J and K take; i.e. both J and K are 'don't cares'. The two additional rows shown in Table 6.6 must be added to Table 6.5.

This addition of 'don't cares' for unused states leads to some problems in a very small number of cases; these problems are examined in Section 6.7.

6.5   Design of the circuit

An excitation table contains all the information required to complete the design of a sequential logic circuit. As stated previously, all the bistable clock inputs must be connected to the same source of clock pulses; this source controls the circuit timing. All that remains is to design circuits which will provide the required J and K inputs for each flip-flop.

Because master-slave flip-flops are used, their Q outputs will not change until the clock goes low (i.e. to 0) but the master section is changed to the correct new output while the clock is high (at 1). Therefore, the existing (present) Q outputs may be used together with the control inputs as inputs of combinational logic circuits whose outputs generate the required J and K inputs. Master-slave devices are essential – otherwise the J and K inputs provided by the combinational circuits will change while the clock is high and might cause incorrect circuit operation.

The truth tables for these combinational logic circuits are contained within the excitation table. For example, the first row of Table 6.5 shows that when A = B = Z = 0, then the four combinational logic circuits which are required must have outputs which generate the required inputs of $J_A = 1$, $K_A = X$, $J_B = 0$ and $K_B = X$. Similarly, the second row indicates that when A = B = 0 and Z = 1, these combinational logic circuits must give outputs which supply $J_A = 0$, $K_A = X$, $J_B = 0$ and $K_B = X$. Therefore, the excitation

Table 6.7

| Present State | | Control | |
B	A	Z	J_A
0	0	0	1
0	0	1	0
0	1	0	X
0	1	1	X
1	0	0	0
1	0	1	0
1	1	0	X
1	1	1	X

table can be used to provide truth tables for combinational logic circuits. There must be one truth table corresponding to the circuit which supplies a single J input, and another for the circuit supplying one K input; i.e. there are two truth tables and two circuits for every bistable. Usually, the excitation table is used directly to give Boolean expressions for the combinational circuits, but to illustrate the procedure fully, the truth table for the circuit whose output is connected to the J_A input in the example is given in full as Table 6.7.

The combinational logic circuit which is required to provide the correct J_A input must obey this truth table, and may be designed using the techniques developed in Chapter 3. The most suitable technique in most cases is to draw the Karnaugh map, and then use it to deduce a minimal Boolean expression for the circuit. Maps for all four circuits required by excitation Table 6.5 are shown in Fig. 6.3. The groups selected on the maps require

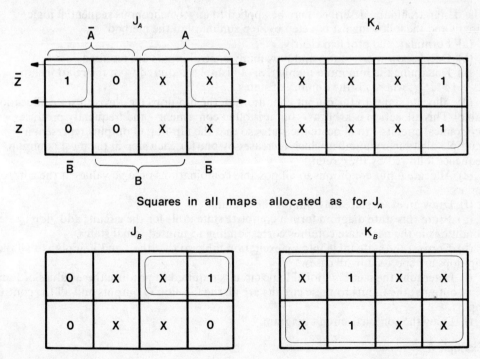

Fig. 6.3.

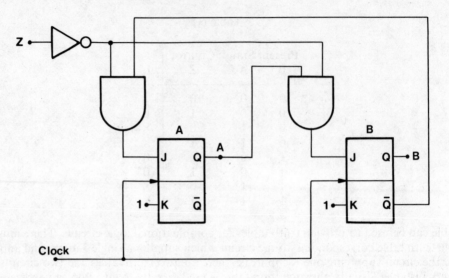

Fig. 6.4.

$J_A = \bar{B}.\bar{Z}$, $K_A = K_B = 1$ and $J_B = A.\bar{Z}$. Thus, the circuits for the J and K control inputs to the flip-flops have been determined. The complete circuit diagram may now be drawn (Fig. 6.4); note the use of the flip-flop $\bar{Q}$ outputs to provide inverted quantities.

6.6 Summary of the design method

The design technique described may be applied to any synchronous sequential logic circuit and the following list is a step by step summary of the method.

(a) Formulate the problem clearly.

(b) Decide how many states and how many control inputs are required.

(c) Determine the minimum number, n, of bistables required from the relationship $2^n \geqslant j > 2^{n-1}$, where j is the number of states.

(d) Allocate states to the output conditions of the flip-flops (i.e. devise a state allocation table). This allocation is arbitrary, but it is often convenient – and frequently produces economical circuits – to allocate the states so that the flip-flop Q outputs represent a multiple digit binary number which increases by one for each step in the most common sequence followed by the circuit.

(e) Allocate input conditions to all possible combinations of logic values at the control inputs.

(f) Draw an exact state diagram for the circuit.

(g) From this state diagram form a complete state table for the circuit; add 'don't care' conditions in the next state columns corresponding to unused initial states.

(h) Convert the state table into an excitation table so that the J and K inputs to all the flip-flops are specified in all cases.

(i) Determine the combinational logic circuits required to produce these values of J and K as outputs; the inputs to these circuits are all the flip-flop Q outputs and all the control inputs.

(j) Draw the complete circuit diagram.

Example 6.3

Design a divide-by-six up counter.

Solution

A detailed solution is given to this problem.

(a) The problem clearly specifies the circuit.

(b) A divide-by-six counter must have six states and as it is a single-direction counter with no special features it has no control inputs.

(c) There are six states hence the number of bistables, n, is given by $2^n \geqslant 6 > 2^{n-1}$ so that n must be three.

(d) Table 6.8 shows the state allocation selected; A, B and C are the Q outputs of the three bistables.

Table 6.8

State	C	B	A
S_1	0	0	0
S_2	0	0	1
S_3	0	1	0
S_4	0	1	1
S_5	1	0	0
S_6	1	0	1

(e) As there are no control inputs there are no input conditions to allocate.

(f) The state diagram is a simple one and is shown in Fig. 6.5.

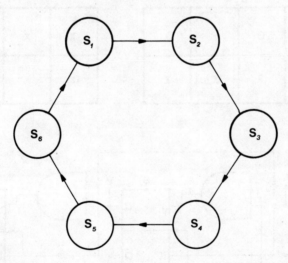

Fig. 6.5.

(g) and (h) From Fig. 6.5 the state table is constructed and the columns required to convert it to an excitation table are added; the result is Table 6.9.

(i) Using Table 6.9, the Karnaugh maps for each J and K input circuit are completed. These maps are shown in Fig. 6.6 and the solutions found are $J_A = K_A = 1$, $J_B = A \cdot C$, $K_B = A$, $J_C = A \cdot B$ and $K_C = A$.

(j) Using these relationships the complete circuit diagram is drawn and is included in Fig. 6.6.

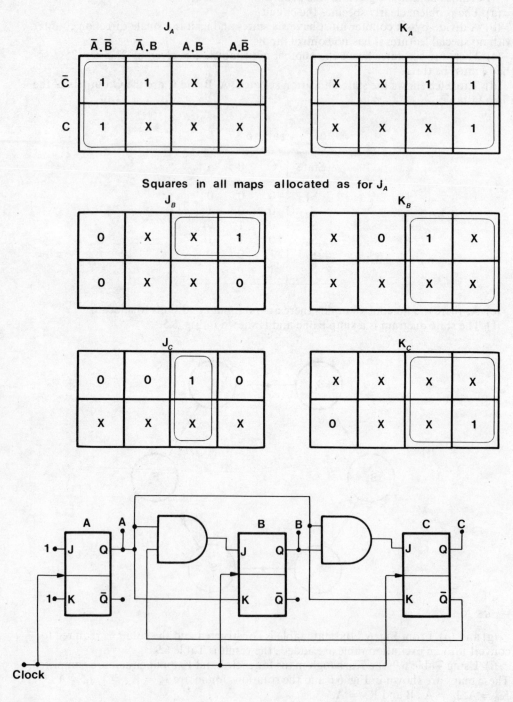

Fig. 6.6.

Table 6.9

Comment	Present State			Next state			Flip-flop controls					
	C	B	A	C	B	A	J_C	K_C	J_B	K_B	J_A	K_A
$S_1 \rightarrow S_2$	0	0	0	0	0	1	0	X	0	X	1	X
$S_2 \rightarrow S_3$	0	0	1	0	1	0	0	X	1	X	X	1
$S_3 \rightarrow S_4$	0	1	0	0	1	1	0	X	X	0	1	X
$S_4 \rightarrow S_5$	0	1	1	1	0	0	1	X	X	1	X	1
$S_5 \rightarrow S_6$	1	0	0	1	0	1	X	0	0	X	1	X
$S_6 \rightarrow S_1$	1	0	1	0	0	0	X	1	0	X	X	1
Unused	1	1	0	X	X	X	X	X	X	X	X	X
Unused	1	1	1	X	X	X	X	X	X	X	X	X

Example 6.4

Design a divide-by-four up–down counter.

Solution

Only a brief outline of the solution is given for this example.

The circuit behaviour is clearly defined in the problem and it is apparent that two bistables are needed to produce the four states required. The method of selecting the count direction has not been specified and may be freely chosen; an obvious choice is to use a single control input to produce two input conditions. Both the state and input condition allocations chosen are given in Table 6.10 while Fig. 6.7 and Table 6.11 show the

Table 6.10

State	B	A
S_1	0	0
S_2	0	1
S_3	1	0
S_4	1	1

a) State allocation

Control input	Input condition	Comment
0	I_1	Count up
1	I_2	Count down

b) Input conditions

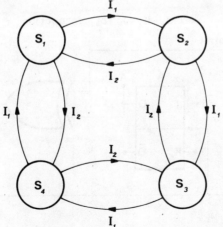

Fig. 6.7.

state diagram and corresponding excitation table respectively. The Karnaugh maps for the circuits which supply the J and K inputs and the final circuit diagram are given in Fig. 6.8.

Table 6.11

| Present state | | Control | Next state | | | | | |
B	A	Z	B	A	J_B	K_B	J_A	K_A
0	0	0	0	1	0	X	1	X
0	1	0	1	0	1	X	X	1
1	0	0	1	1	X	0	1	X
1	1	0	0	0	X	1	X	1
0	0	1	1	1	1	X	1	X
0	1	1	0	0	0	X	X	1
1	0	1	0	1	X	1	1	X
1	1	1	1	0	X	0	X	1

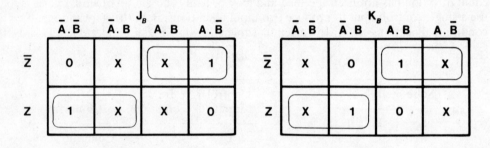

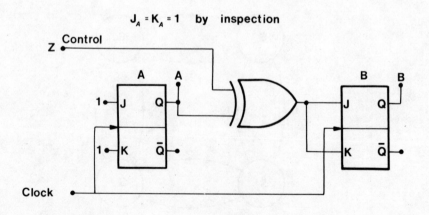

Fig. 6.8.

6.7 Problems associated with unused states

It was indicated in Section 6.4 that operating problems may arise in circuits which have been designed with 'don't care' conditions in the next state column of the state table for cases when the initial state is an unused (redundant) one. The problems arise because at switch on each bistable settles at random with either possible value of Q output and the circuit takes up any possible state including the unused ones.

Often there is no problem as a few clock pulses drive the circuit from any unused state into a used one and thereafter the circuit follows the correct sequence. However, in some cases the circuit may become trapped in an unused state or group of unused states. Three different state diagrams for a divide-by-five up counter with three unused states S_6, S_7 and S_8 are shown in Fig. 6.9; the unused states are shown in three of the many possible ways.

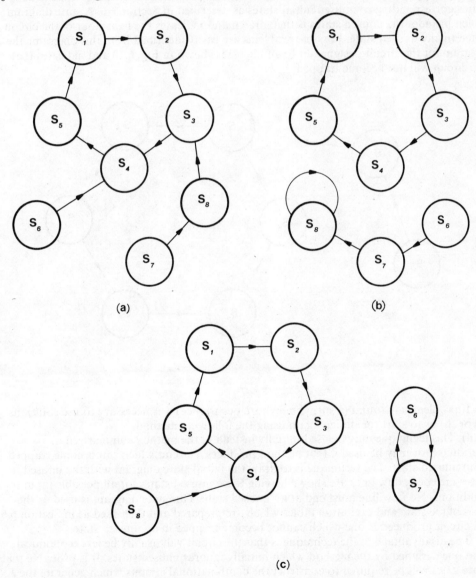

Fig. 6.9. Some state diagrams for a divide-by-five counter

In case (a) the circuit cannot become trapped in an unused state but in the other two cases it may be trapped.

Several methods are available to avoid or overcome this particular problem.

(a) After power is switched on, a manual reset switch is used; this is connected so that the circuit is forced into a known used state. This method, or the next one, is commonly used with very large systems and has the additional advantage that the circuit starts at a known position in its sequence.

(b) Method (a) is automated; an electronic or electromechanical device senses the rise in supply voltage at switch on and provides the forced reset when the voltage reaches the correct operating level.

(c) The problem is ignored and the circuit is designed using 'don't cares' for the next states corresponding to unused initial states as described in Section 6.6. A state diagram which includes the unused states is then drawn and indicates the behaviour of the circuit when in any unused state. Possible problems are easily identified from this diagram; the diagram for the circuit designed in Example 6.3 is shown in Fig. 6.10 and indicates that the circuit will not become trapped.

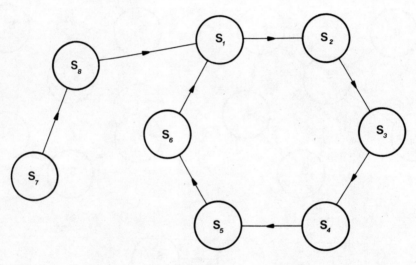

Fig. 6.10.

If problems are found (a surprisingly rare occurrence) it is necessary to use solutions (a) or (b) or to start the design again using the following method.

(d) This technique may be used initially instead of the method summarized in Section 6.6, or may be used if that method produces a circuit which can become trapped in an unused state. The technique is to draw the initial state diagram with the unused states included; flow lines are shown leaving these unused states for all possible input conditions. No flow line must end at an unused state. This state diagram is used as the basis of the state and excitation tables which are prepared and then used as in Section 6.6; the circuit produced is one which cannot become trapped in an unused state.

The disadvantage of the technique is that the circuit will usually be less economical than one designed by the method which initially ignores unused states. It is probable that more gates will be required to construct the combinational circuits which generate the J and K inputs.

Example 6.5

Design a synchronous divide-by-six up counter arranged so that any transitions from unused states end at the first used state.

Solution

This is Example 6.3 repeated with the modification that unused states are not neglected. Steps (a) to (e) in the design process are identical but at step (f) a revised state diagram is required and is shown in Fig. 6.11. The new excitation table is easily derived; it

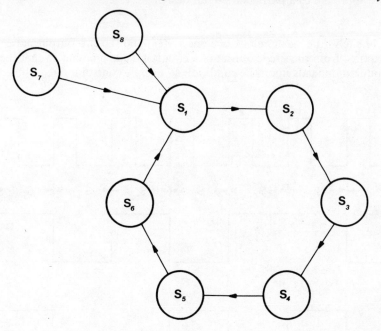

Fig. 6.11.

is Table 6.12 and the expressions derived from it for the connections to the J and K inputs are $J_A = \bar{B} + \bar{C}$, $K_A = 1$, $J_B = A . \bar{C}$, $K_B = A + C$, $J_C = A . B$ and $K_C = A + B$. Some of these expressions are obviously more complex than the corresponding ones obtained in Example 6.3. However, any circuit constructed using these expressions will not stick an unused state; its behaviour is completely specified and no further checks are required.

Table 6.12

Present state			Next state			Flip-flop controls					
C	B	A	C	B	A	J_C	K_C	J_B	K_B	J_A	K_A
0	0	0	0	0	1	0	X	0	X	1	X
0	0	1	0	1	0	0	X	1	X	X	1
0	1	0	0	1	1	0	X	X	0	1	X
0	1	1	1	0	0	1	X	X	1	X	1
1	0	0	1	0	1	X	0	0	X	1	X
1	0	1	0	0	0	X	1	0	X	X	1
1	1	0	0	0	0	X	1	X	1	0	X
1	1	1	0	0	0	X	1	X	1	X	1

6.8   Design from timing diagrams

So far, the sequential circuit designs examined have been produced to meet a written specification. When a large system is designed by a team an individual designer may be required to produce a sub-system for which the specification is a timing diagram. This is because in a large system there is usually a master clock (often multiphase) which produces a continuous clock pulse train that controls the time at which all changes occur.

The range of timing diagrams for possible circuits is infinite; the following example illustrates one approach to a particular circuit design.

Example 6.6

Figure 6.12 shows the master clock of some system and the two output pulse sequences required from the single output of a circuit. Design a circuit so that the output sequence produced depends upon the condition at a single control input.

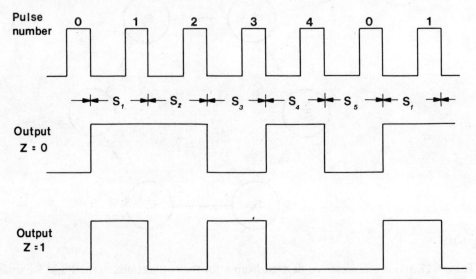

Fig. 6.12.

Solution

This is a simple design problem as both output sequences in Fig. 6.12 have the same cycle length (repetition interval). This cycle length can be determined in terms of a number of clock pulses – five in this example – and this number is also the number of states (N) required in the sequential circuit. Once the cycle length is known, a divide-by-N single direction counter is designed using any standard technique. In cases for which variable length sequences are required, the counter must have control inputs which are connected so that the cycle length depends upon the input conditions.

The final output is produced by a combinational logic circuit which has the control input (or inputs) and the bistable Q outputs as its inputs. This is the type of output circuit which was briefly introduced in Section 6.1c. For the example let states S_1, S_2, S_3, S_4 and S_5 correspond to the intervals which occur after clock pulses 0, 1, 2, 3 and 4 respectively. The output circuit truth table is constructed and shows the final circuit output required from the combinational circuit for all possible combinations of input conditions and bistable Q outputs. Unused states are considered as never existing and the final output corresponding to such states may be taken as 'don't care'.

Table 6.13

State	Counter flip-flop Outputs			Control	Circuit
	C	B	A	Z	Output
S_1	0	0	0	0	1
S_1	0	0	0	1	1
S_2	0	0	1	0	1
S_2	0	0	1	1	0
S_3	0	1	0	0	0
S_3	0	1	0	1	1
S_4	0	1	1	0	1
S_4	0	1	1	1	0
S_5	1	0	0	0	0
S_5	1	0	0	1	0
Unused	1	0	1	0	X
Unused	1	0	1	1	X
Unused	1	1	0	0	X
Unused	1	1	0	1	X
Unused	1	1	1	0	X
Unused	1	1	1	1	X

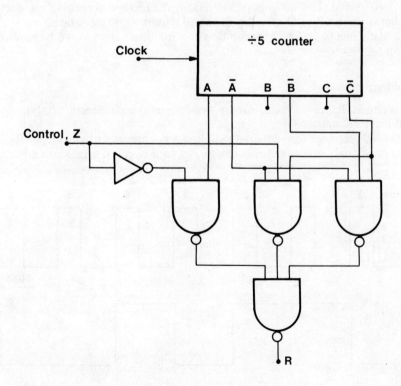

Fig. 6.13.

Table 6.13 is the output circuit truth table for the example and may be used to produce a minimal logic expression for the circuit output. The result obtained is

$$R = A.\bar{Z} + \bar{A}.\bar{B}.\bar{C} + \bar{A}.\bar{C}.Z = \overline{\overline{(A.\bar{Z})}.\overline{(\bar{A}.\bar{B}.\bar{C})}.\overline{(\bar{A}.\bar{C}.Z)}}$$

When a combinational circuit which implements this is connected to the divide-by-five counter the circuit in Fig. 6.13 is obtained.

6.9   Comments

This treatment of sequential circuit design provides a relatively simple technique; the main disadvantage is that at no stage has the restriction been imposed that the most economical circuit should be produced. Some more advanced techniques attempt to minimize the circuit but can rarely take into account all possible reductions.

For example, a simple divide-by-four counter requires four states and the method described here allows the states to be allocated to the values of the bistable Q outputs in an entirely arbitrary manner. It was suggested that a systematic allocation be used but this is not essential; in a simple four-state circuit there are twenty-four different ways in which the allocation could be made. Any more advanced technique should determine the allocation which leads to the most economical solution. However, as electronic components are inexpensive and very reliable this non-minimal solution is not a major disadvantage when the technique described is used for electronic systems. It is more important to produce a hazard-free circuit than to impose any other limitation on the design of a sequential circuit.

The design technique may be adapted to utilize other types of master-slave flip-flops. All that is necessary is to derive the excitation table (equivalent to Table 6.4 obtained for the JK flip-flop) for the chosen flip-flop then modify the method accordingly. Generally circuits designed using SR or D-type flip-flops will require more elaborate combinational circuits to generate the flip-flop control inputs than would be required for circuits using JK flip-flops.

6.10   Problems

1 Produce a circuit diagram for a divide-by-five counter (single direction only) constructed using JK master-slave flip-flops.
2 Design a two-directional (up–down) divide-by-six counter constructed from JK master-slave flip-flops. Determine the behaviour of your circuit if it gets into any unused state.

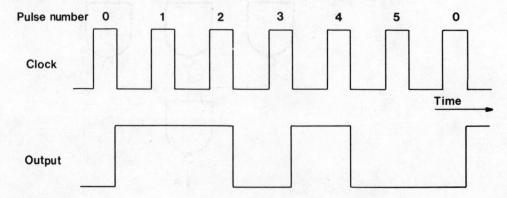

Fig. 6.14

3 Design a single-direction counter using three JK master-slave flip-flops. The counter must have five states (i.e. it must be divide-by-five up counter) but the counting sequence must be such that if the three Q outputs are taken as the digits of a BCD number, the count sequence is 0, 1, 3, 5, 6, 0, 1, 3 etc.

4 Derive the circuit diagram of a synchronous counter with a single control input; when the control input is 1 the counter behaves as a divide-by-seven up counter and when the control input is 0 the behaviour is a divide-by-five up counter.

5 Figure 6.14 is a timing diagram showing the clock input and single output of a sequential logic circuit. Devise a circuit which behaves as required by this diagram.

6 Devise a sequential logic circuit to control traffic lights (traffic signals) at a simple crossroads junction. Assume that a clock pulse is provided every two seconds and that while changing, the lights remain at an intermediate state for two seconds. The lights operate on a simple time control sequence (no traffic-actuated control); traffic on each road is allowed to move in turn and the lights remain red or green for ten seconds.

7 Design a divide-by-four up–down counter using SR master-slave flip-flops.

7 Electronic Logic Circuits

In the preceding chapters it was usually assumed that logic circuits could be constructed using ideal components whose internal construction imposed no limitations on circuit design. Under such conditions the only restrictions which apply to the design of logic networks are those set by the laws of Boolean arithmetic and algebra. Any real system will be further restricted because it must be built with components with a performance which is not ideal and designers must make allowances for component limitations.

At present most students of electronic logic design and construct circuits using 7400 series TTL components (5400 series components are more robust equivalents) or 4000 series CMOS components and detailed comments are restricted to these families of devices. A partial explanation of manufacturers' code numbers is included later in Table 7.2. Although comments refer only to these two types of logic devices, other electronic and many non-electronic logic elements will impose similar constraints upon network operation and design. The reasons for the non-ideal behaviour of electronic logic elements require detailed consideration of the internal electronic circuit construction; this topic is not examined here. Numerical values for the limiting parameters of individual devices may be obtained from the manufacturers' published data.

In addition to elements (gates, flip-flops etc.) which may be described in terms of Boolean functions, most families of electronic logic devices include special components which are not strictly two-state logic devices. These components allow certain problems to be solved in a more economical manner than is possible if only true two-state devices are available; the most common special components are open collector and tri-state output devices.

It is convenient to divide the various problems which arise when designing circuits using available components into groups within which the causes and solutions of problems are similar.

7.1 Timing problems

Some circuit problems arise because real logic elements take a finite time to operate **(propagation delay)** and one or two of these problems have already been described. Timing hazards are an inherent feature of all practical logic systems and two hazards have already been examined. These are the race condition (or critical race) described in Section 4.1 and the production of transient outputs by asynchronous circuits.

7.1.1 Race conditions

Strictly speaking, a race condition does not arise because logic elements exhibit propagation delay. A problem in predicting circuit operation exists whenever two signals simultaneously attempt to produce opposite effects and this problem would be present even if the elements had zero propagation delay. The effect of finite, but marginally different, propagation delays often allows the system to settle into an unknown but stable state as in the case of the SR flip-flop with $S = R = 1$ examined in Section 4.1. In other cases the circuit may oscillate and Fig. 7.1 shows an example of one such circuit; it consists of five inverters connected in a single closed loop and a simple investigation

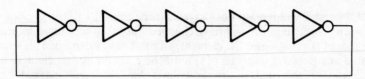

Fig. 7.1. Five-inverter loop oscillator

illustrates the race condition. Assume that a particular logic level is present at the input to one gate and determine the conditions at each output around the loop; the analysis indicates that the opposite level to the one assumed should be present at the original input. Because the elements have a finite propagation delay the circuit oscillates with a period of oscillation which is approximately ten times the mean propagation delay for a single gate.

If perfect (i.e. zero propagation delay) elements were used to construct an SR flip-flop and a five-inverter loop, the exact circuit behaviour for the race conditions could only be determined with knowledge of the internal construction of the elements. It is possible that in such circumstances the circuit would settle into a condition with the outputs outside the range defined for the logic levels.

The race conditions are obvious in these two simple examples but it is often very difficult to identify possible race conditions in large networks. The most successful method of identifying race conditions and other hazards is to program a computer to simulate the network; the program output is such that it indicates possible problems. When the design techniques described in Chapter 6 are correctly applied, race conditions should not arise.

7.1.2 Transient outputs

The production of transient outputs by asynchronous circuits is a major problem associated with such circuits. Transients of very short duration may even arise in well-designed synchronous circuits if the circuit components have widely differing propagation delays. However, transient problems with synchronous circuits are uncommon if all the components within the network are chosen from a single logic family.

Transient outputs may also be produced by combinational networks. The circuit of Fig. 7.2 illustrates this; it implements the function

$$R = A.\bar{B} + \bar{A}.C$$

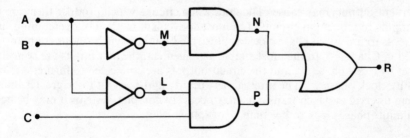

Fig. 7.2.

Suppose that at some instant in time, t_0, the circuit input conditions change instantaneously from A = 1, B = 0, C = 1 to A = 0, B = 0, C = 1; i.e. A changes. Assume that all the logic elements have an identical propagation delay of t_d then at times between t_0 and $t_0 + t_d$ input A will have changed but the output, R, and all the intermediate points marked L, M, N and P will remain unchanged. At the time $t_0 + t_d$, those gates which are connected directly to the input A will have changed but those connected indirectly (i.e. one or more gates between A and the element) will not have changed. This examination can be repeated at successive intervals of t_d until there are no further changes in any gate outputs. Full details of this analysis are shown in Table 7.1; entries marked with an asterisk indicate cases for which a gate output has not had time to change and the output is not the one implied by the Boolean function for the gate with the input conditions shown.

Table 7.1

Time, t, from $t_0 = 0$	A	B	C	$L = \bar{A}$	$M = \bar{B}$	$N = A.M$	$P = L.C$	$R = N + P$	Comments
Before t_0	1	0	1	0	1	1	0	1	Initial conditions
$t_d > t > 0$	0	0	1	0*	1	1*	0	1	New input, L and N must change
$2t_d > t > t_d$	0	0	1	1	1	0	0*	1*	L and N have changed, this requires that P and R change.
$3t_d > t > 2t_d$	0	0	1	1	1	0	1	0*	P and R have changed but the new value of P requires R to change again.
After $t = 3t_d$	0	0	1	1	1	0	1	1	Final condition

Examination of the column for the output R in Table 7.1 indicates that although both the initial and final input conditions give an output of 1, a different output of 0 is produced for an interval of t_d. Thus a transient output is produced by a combinational logic circuit; such transients and more complex ones (e.g. a sequence 0–1–0–1 for a simple 0 to 1 change) are quite common in very large combinational networks. Removal of the restriction that all gates have identical propagation delays complicates the analysis, but does not significantly alter the results.

Design techniques exist which produce 'transient free' combinational circuits, but they only work if circuits are constructed using gates selected with approximately the same propagation delays. Such 'transient free' circuits have the additional disadvantage that they are often less economical than circuits designed ignoring the possible production of transients.

Transient outputs may cause difficulties when circuits which produce them are used to supply inputs to sequential circuits. The most convenient method of overcoming such difficulties is to use a two phase clock as illustrated by the asynchronous counter described in Chapter 5. In the most simple arrangement, the new inputs are connected at a time set by clock phase ϕ_1, and the circuit outputs are connected to further circuits at a time set by clock phase ϕ_2. The interval between ϕ_1 and ϕ_2 is set to be greater than the maximum interval in which transients may occur. In complex systems it may be necessary to use a multi-phase clock rather than a two-phase one.

7.2   Interconnections

One assumption made previously is that the output of one logic element is capable of being connected to supply the inputs to any number of other elements. In a real system (electronic or other) the output of one element can only drive a finite number of inputs to further elements; this finite number is termed the **fan-out** of the element. It is usual to ensure that the standard elements in any family of logic devices have a fixed input drive

Table 7.2

Series	Code	Characteristics	Fan-in	Fan-out	Typical single gate propagation delay (D)	Typical single gate power consumption (P)	Speed-power product (D × P)
Standard	p74XXs	Basic series	1	10	10 nsec	10 mW	100 pJ
Low power	p74LXXs	Lower power consumption but slower	0·25	2·5	33 nsec	1 mW	33 pJ
High speed	p74HXXs	High speed operation, requires high power.	1·25	12·5	6 nsec	22 mW	132 pJ
Schottky	p74SXXs	Schottky clamped device. Faster than H series.	1·25	12·5	3 nsec	19 mW	57 pJ
Low power Schottky	p74LSXXs	Low power version of S series. Approximately standard series speed with L series power.	0·25	5·0	9·5 nsec	2 mW	19 pJ

Key to device codes:
   p. A prefix unique to a manufacturer. (e.g. N for Signetics, SN for Texas Instruments, MC for Motorola).
   s. A suffix which may vary from one manufacturer to another; it specifies the package (e.g. plastic or ceramic) and other mechanical features.
   XX. A two or three digit number which specifies the device function (e.g. 10 is triple 3-input NAND).
Examples:
   N74LS04N is a Signetics, Low Power Schottky, hextuple inverter in a plastic package.
   SN7420N is a Texas Instruments, standard series, dual 4-input NAND gate in a plastic package.

requirement which may be defined as one unit. The fan-out capability of any element may be given in terms of these units; the standard fan-out of 7400 series TTL devices is ten.

As logic systems are developed, it becomes necessary to design components with non-standard input drive requirements. The input drive requirement is termed the **fan-in** and is also in units of the standard input requirement. For example, many 7400 series JK master-slave flip-flops have J and K inputs with a fan-in of one (i.e. they are standard) but the fan-in for the clock input is two. Thus, a standard gate may be used to drive up to ten J and K inputs, but is only capable of driving up to five clock inputs connected in parallel.

In addition to devices with fan-in ratings which are not standard there are devices with non-standard fan-out ratings. For example, devices with type numbers 7440 are dual four-input NAND gate buffers; i.e. they consist of two separate four-input NAND gates, and each gate has a fan-out of thirty. The situation is further complicated because the popularity of 7400 devices has led to the introduction of several modified series, each with special properties. These devices, their special properties and the typical fan-in and fan-out for such devices are summarized in Table 7.2. In all cases, the units used for fan-in are the units for devices in the standard series.

The various types of 7400 components may be interconnected but it is essential to ensure that the output of a device is capable of driving all the inputs connected to it. Table 7.2 is only a summary and indicates worst-case situations. Interconnections designed within the limits given will operate reliably. In certain applications, less conservative ratings may be used, and may be determined by careful study of component manufacturers' specifications.

Example 7.1

Determine whether the following connections will operate reliably.

(a) A standard 7400 series device driving seven Schottky (74S) devices.

(b) A low power (74L) device driving three standard inputs.

(c) A low power Schottky (74LS) device driving one standard and two high speed (74H) inputs.

Solution

(a) Each 74S input has a fan-in of 1·25 hence the total drive requirement is $7 \times 1·25 = 8·75$. As the standard device has a fan-out of ten the connection may be made.

(b) The standard inputs have a fan-in of 1 and the total load is 3; however, the fan-out of most 74L devices is only 2·5 so that it is incapable of driving the inputs to three standard devices. The connection should not be made, although 2·5 is so close to 3 that a circuit constructed with this connection might operate successfully.

(c) The load is one standard device, fan-in 1, plus two 74H devices each with fan-in 1·25, which gives a total of $1 + 2 \times 1·25 = 3·5$. As the 74LS element used has a fan-out of 5, the connection will operate reliably.

Note that in all the situations examined in this example it has been assumed that the elements used have the typical fan-in and fan-out for their own series.

Interconnection rules for 4000 series CMOS devices are not quite so simple as those for 7400 series TTL devices. In theory a CMOS output is capable of driving an infinite number of further CMOS inputs, but in practice there is a finite, although large, limit. This limit cannot be described by a single number but must be determined for each application because the fan-out is determined by two operating conditions selected by the circuit designer. These conditions are the supply voltage (not fixed for CMOS unlike TTL) and the propagation delay; their relationship to fan-out is illustrated by Fig. 7.3.

Hence, to obtain a numerical value for the fan-out of a CMOS element, the supply voltage and maximum tolerable propagation delay must be specified. The fan-out is then

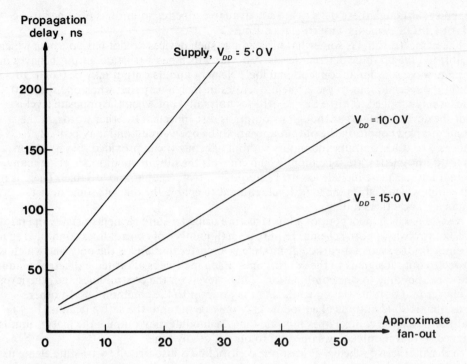

Fig. 7.3. CMOS element fan-out

obtained by examination of Fig. 7.3; this value should be slightly reduced (typically by about 10%) to allow for effects caused by the interconnection leads. Generally, fan-outs above one hundred should be avoided.

Note that two circuit outputs must *never* be connected together; Boolean algebra does not define the result of such a connection and electronic logic elements are not designed to allow these connections. If such a connection is made the resulting output will be indeterminate, and the devices may be permanently damaged.

7.3 Special interconnection techniques

It is convenient to assemble large systems from a number of separate independent modules or units, with the modules linked together by a large number of parallel connections. Usually modules may be plugged into, or removed from, this connection system; often they may be positioned at any point on the system and may use the parallel connections to send an output to, or receive an input from, another module.

This assembly technique has the advantage that one-off quantities of special-purpose systems may be manufactured from standard units; each unit is therefore inexpensive and well tested. In addition, any fault may be localized by replacing one module at a time until the fault is traced to a single unit.

The arrangement of parallel links between modules is termed a **bus** or a **highway** or a **dataway**. Obviously, the rule that outputs must not be connected together will be broken by the bus arrangement; also, the fan-out capability of an element may be exceeded since it is outside the control of the designer of the modules. There are techniques which allow a bus system to be implemented using standard components, but the most satisfactory method is to use special devices. One such device is the TTL gate with **'wired-OR'** capability; although these devices are still available a better solution is obtained using TTL or CMOS elements with **tri-state** outputs.

The name 'tri-state' is somewhat misleading; such devices do not have outputs which are able to take up three different levels. A device which has a tri-state output behaves in the same way as a similar conventional logic element and its output may be 0 or 1. In addition, a special input to the element operates in such a way that when one specified logic level is supplied to it the device behaves normally, but when the opposite level is input the device operates as though its output is disconnected. In other words, in the second disabled condition the output appears to be open circuit and may be freely connected to other outputs and inputs without affecting them, provided that it remains in this third (disabled) state. The input which controls the output condition is given many different names; these include 'control', 'output control', 'enable' and 'disable'. There is no convection which defines the logic level required to enable the output; some devices require a 1 and others a 0.

A system which incorporates a bus structure using tri-state elements usually operates in one of two ways; both schemes require a small number of control lines connected to all the units. In one system a particular module is a master unit and is the only one which is allowed to output signals to the control lines. Each module is allocated a different unique code corresponding to one combination of logic levels on the control lines and whenever the master unit outputs this code the unit may output to the common bus. In other words, the tri-state outputs of a module may only be put into the active (enabled) condition when the control lines indicate that the module may output; the master unit is designed so that it only allows one unit to output at any time.

In the alternative scheme, at least one control line is also driven by tri-state elements of every unit. In the most simple arrangement this line is connected to the logic 1 level through a resistor (pull up resistor). All units are of equal status and may output to the bus at any time, provided that the control line is at 1. To output to the bus a module examines the control line and if it is at 1 it enables its own tri-state output to the control line and forces the line to 0. Once a particular module has forced the control line to 0 it may output to the bus and hence to other units. When the output operation is complete the module disables all its tri-state outputs, including that to the control line, and the control line returns to 1. If a module attempts to output while another unit is using the bus it will find that the control line is 0 and must wait until the line becomes 1 before it can output.

Both techniques are usually implemented in a slightly more complex form to overcome problems which arise if two or more units request access to the bus simultaneously. In addition, more advanced systems may include methods by which a unit sending signals to another unit may check that they have been correctly received.

7.4   External connections

Electronic circuits, no matter how large or sophisticated, are of no use by themselves. An electronic circuit will be used to perform some task and will frequently involve connections to non-electronic components. For instance, a calculating device must have some method of receiving details of the calculations to be performed and some way of indicating the results. In the most simple case the input will be from a keyboard (i.e. a set of switches) and the result will be indicated by a set of lamps. A more elaborate electronic logic device such as a controller for a machine tool may have to supply inputs to drive motors, relays, solenoids, etc., all of which must be operated by the electronic circuits.

In general, the input and output circuits (interface circuits) which connect an electronic logic circuit to external devices or to linear electronic circuits (i.e. non-logic ones) require a detailed knowledge of electronic circuit construction for their design. In a few cases integrated circuits are available for particular interface tasks, but generally, interface design requires experience – although one or two simple interfacing problems may be solved using the circuits below.

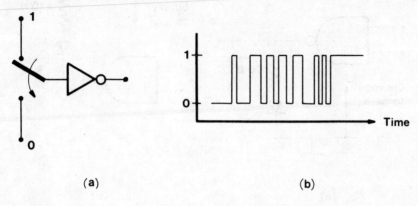

(a)   (b)

Fig. 7.4.

It might appear to be reasonable to use the circuit in Fig. 7.4a to provide an input to a logic element (an inverter is illustrated) from a manually-operated switch. However, this is not a satisfactory circuit because every time a mechanical switch is operated, the spring material from which it is manufactured vibrates and the contacts open and close several times instead of making a single contact. The result of this multiple action is that the output of the logic element in Fig. 7.4a is similar to the waveform shown in Fig. 7.4b. This switch 'contact bounce' lasts for several milliseconds even with high quality switches, and in many applications the effects must be removed.

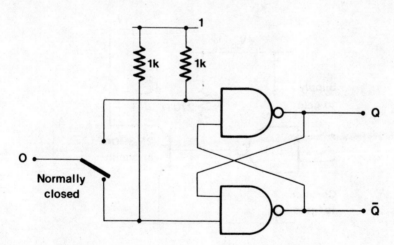

Fig. 7.5. Switch buffer circuit

Figure 7.5 shows one circuit which removes the problem, it is basically a $\bar{S}\bar{R}$ flip-flop with the $\bar{S}$ and $\bar{R}$ inputs connected to a spring-loaded changeover switch. The switch must be constructed such that one set of contacts open before the other set closes (break before make) and a single press and release operation produces a single 0 to 1 to 0 pulse at Q. There is a simultaneous 1 to 0 to 1 pulse at the $\bar{Q}$ output; in this application the circuit is termed a switch buffer circuit.

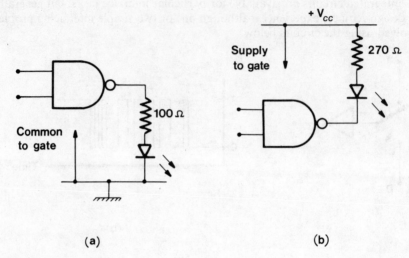

(a)   (b)

Fig. 7.6. TTL to LED drive circuits

Output circuits are more difficult; for simple indications of logic output levels a light-emitting diode (LED) may be used with TTL circuits as shown in Fig. 7.6. Using the circuit in Fig. 7.6a, the LED is on (illuminated) when the output of the gate is 1 and it is off when the output is 0; using the circuit in Fig. 7.6b, the reverse is the case.

Another simple output circuit is one which will provide the drive current to the coil of a relay and one such circuit is shown in Fig. 7.7. When the logic element gives an output

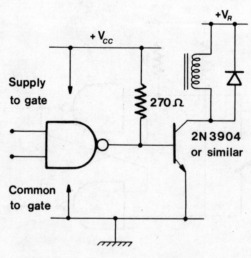

Fig. 7.7. TTL to relay drive circuit

of 1, the relay operates (current flows in the coil) and when the output is 0, the relay is inactive. The relay coil must have an operating voltage of V_R, usually between 5 volts and 15 volts, and the operating current should not exceed about 50 mA. It is essential to ensure that the relay contacts driven by the coil are adequately rated for the task to which they are applied.

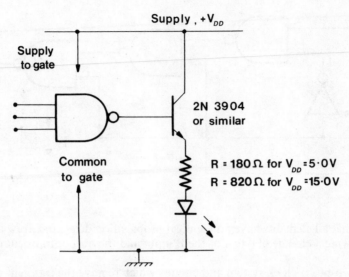

Fig. 7.8. CMOS to LED drive circuit

The LED and relay drive output circuits should not be used with CMOS elements because their output is not capable of providing sufficient current. A simple output circuit driving a LED display from a CMOS device is shown in Fig. 7.8.

7.5 General assembly points

Circuit connections should be short and neat; whenever possible, leads should be well separated and if possible an earthed lead should be placed between adjacent signal leads. For simple systems which are not required to operate at maximum possible speed reasonable care will produce a reliable circuit. Large assemblies can only be constructed satisfactorily using printed circuit board (PCB) techniques; these must be used with high speed circuits and in such applications circuit layout requires knowledge of signal transmission techniques.

In addition to the power supply producing a steady output voltage at the required level, it is necessary to connect capacitors of about $0.1\,\mu F$ between the power supply leads close to each group of four or five integrated circuits. These capacitors should be a high-frequency type and are required because the internal construction of TTL and CMOS elements is such that large currents (relative to those normally required) flow when the output changes state. For TTL elements the currents are typically above $20\,mA$ and of about 10 nanoseconds duration. These currents generate voltage spikes on the power supply lines; the capacitors remove these spikes and prevent one circuit element interfering with another through the power supply leads.

The final assembly point to consider is the manner of dealing with the unused inputs which arise in some circuit designs. These must always be connected to a suitable logic level and not left open circuit. This logic level may be one of the power supply lines but TTL inputs should be connected to a 1 level through a $1\,k\Omega$ resistor rather than directly connected.

7.6 Problems

1 Demonstrate that the circuit shown in Fig. 7.9 will generate a transient output when the inputs change from A = 1, B = 0, C = 1 to A = 1, B = 1, C = 1.

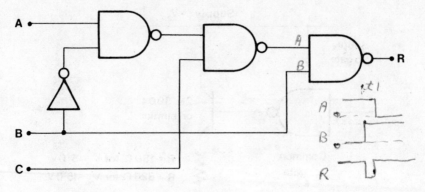

Fig. 7.9.

Assume that all elements have an identical propagation delay and draw a timing diagram showing a change of 0 to 1 at the B input and the output transient on a common time-scale.

2 Devise a two-phase clock system and circuits which remove the transient output produced by the circuit in Fig. 7.9 examined in Problem 1.

3 Determine if the following conditions will operate reliably.

(a) A standard 7400 series output driving two Schottky 74S series inputs and four standard devices.

(b) A low power 74L series device driving two high speed 74H series inputs.

(c) A standard 7400 series output connected to five 74LS series devices and two 74L series devices.

(d) A low power Schottky 74LS output driving four standard 7400 series inputs and two 74S series inputs.

4 Design a system utilizing tri-state output devices. The system has two control lines which provide four different codes and a third line used for timing. These control lines define the action of a circuit which consists of four lines carrying logic signals connected to a single transmission line. The two code control lines identify one of the signal lines according to the code, and when the third line is in the logic 1 state, the line selected is connected to the transmission line so that signals on the selected line are transmitted.

8 Examples and Applications of Logic Circuits

Logic circuits are used in a wide range of products manufactured for domestic and industrial use. The following is an arbitrary selection of a few of the most simple applications, some alternative approaches to particular design problems, and a brief description of connections between logic and non-logic systems.

8.1 Addition of binary numbers

One example of the addition of two binary numbers is illustrated in Table 8.1a and the general case of the ith column of such an addition is shown in Table 8.1b.

Table 8.1a

Augend	1	1	1	1	0	0	1	1
Addend	0	1	1	0	0	1	1	0
Carry-in	1	1	0	0	1	1	0	—
Carry-out	1	1	1	0	0	1	1	0
Sum	1	0	1	0	1	1	0	0	1

Table 8.1b

		Column i	
Augend		A_i	
Addend		B_i	
Carry-in		C_{i-1}	
Sum	C_i	S_i	Carry generated in column $i-1$

Sum in column i

Carry to column $i+1$

8.1.1 Single digit addition

When two binary numbers A, the **augend**, and B, the **addend**, are added together Table 8.1b indicates that the result in the ith column will be a sum of S_i (0 or 1) in this column and a carry-out of C_i (also 0 or 1) to be added to the $i+1$th column. This result represents the arithmetic (not the Boolean) sum of the digits A_i, B_i and C_{i-1} where A_i and B_i are the ith digits of A and B respectively and C_{i-1} is the carry-in which is generated as the carry-out from the previous $i-1$th column. If these binary digits are represented by Boolean variables and the addition is performed by a logic circuit which has inputs A_i, B_i, C_{i-1} and produces outputs of S_i and C_i, then Table 8.2 is the circuit truth table.

Table 8.2

Inputs			Outputs	
Carry-in	Digits		Sum	Carry
C_{i-1}	B_i	A_i	S_i	C_i
0	0	0	0	0
0	0	1	1	0
0	1	0	1	0
0	1	1	0	1
1	0	0	1	0
1	0	1	0	1
1	1	0	0	1
1	1	1	1	1

A Karnaugh map, or similar technique, could be used to derive Boolean expressions for S_i and C_i in terms of the inputs. However, the sum of products solution produced by such a technique is not the one which is most conveniently implemented in many cases; an alternative solution is obtained by the following two-stage treatment. Consider first the case in which the carry-in from the previous stage is not connected; i.e. only A_i and B_i are inputs to the addition circuit. Such a circuit is a **half-adder** and its outputs may be denoted by S_{i0} and C_{i0} where the subscript 0 indicates that there is no carry-in. Table 8.3 is the truth table for the half-adder and inspection gives the relationship

$$S_{i0} = A_i \cdot \bar{B}_i + \bar{A}_i \cdot B_i = A_i \oplus B_i$$

$$C_{i0} = A_i \cdot B_i$$

Table 8.3

Inputs		Outputs	
B_i	A_i	S_{i0}	C_{i0}
0	0	0	0
0	1	1	0
1	0	1	0
1	1	0	1

Figure 8.1 is the circuit diagram of a half-adder constructed from NAND gates and has S_{i0} and C_{i0} as outputs. The name half-adder is used because two half-adders may be connected to form a circuit which performs the complete addition with the carry-in from the previous stage connected. This complete circuit is a **full-adder** and may be described in terms of the half-adder outputs by the relationships

$$S_i = S_{i0} \oplus C_{i-1}$$

$$C_i = A_i \cdot B_i + S_{i0} \cdot C_{i-1} = C_{i0} + S_{i0} \cdot C_{i-1} = \overline{\bar{C}_{i0} \cdot (S_{i0} \cdot C_{i-1})}$$

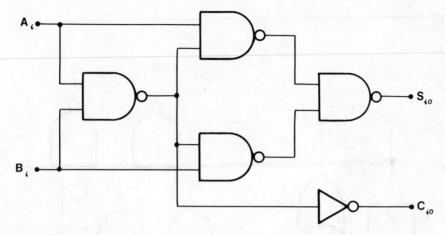

Fig. 8.1. Half-adder circuit

The circuit diagram for a full-adder constructed from half-adders and an expanded diagram showing the half-adders implemented using NAND gates are both shown in Fig. 8.2.

8.1.2 Multiple digit addition

A single full-adder only produces the result of the addition in a single column (i.e. the sum of one pair of digits and the carry-in) when generating the sum of two multiple-digit numbers. To add all the digits there are two possible approaches; either a single full-adder may be used and the digit pairs added one after the other starting with the least significant pair, or the addition unit may have a separate full-adder for each pair of digits.

The technique which uses a single full-adder is known as **serial addition**, a five-digit version is shown schematically in Fig. 8.3. In this design each number is stored in a serial shift register, the carry-out is stored by a D-type master-slave flip-flop, and a control circuit ensures correct operation. The illustration shows no details of the control circuit and the augend register is used to store the result so that the original augend is 'lost'.

When the addition is performed by a circuit with a separate full-adder for each digit pair the circuit is a **parallel addition** one; a five-digit version is shown in Fig. 8.4. In such a circuit all the digits are input simultaneously, often by storing them in parallel in-parallel out (PIPO) registers, and the carry-out from one full-adder is connected as the carry-in to the next. The result appears at the sum outputs of all the adders.

Obviously a serial addition circuit is more economical than a parallel one in terms of the number of logic elements required but it will operate more slowly; the control circuit of the serial adder is complicated. One fault of the parallel adder is the successive carry-out to carry-in connections; these connections generate long propagation delays. More elaborate circuits exist which overcome these long delays at the expense of added circuit complexity.

The development of large scale integrated (LSI) circuits enables large logic networks, for example many digit parallel adders, to be constructed cheaply and to operate reliably. Serial addition circuits are now uncommon; their main application is in the arithmetic unit of pocket calculators which do not operate in a simple binary mode.

All other arithmetic circuits (i.e. subtractors, multipliers, etc.) may be based on addition circuits although most medium and large computers have special-purpose multiplication circuits.

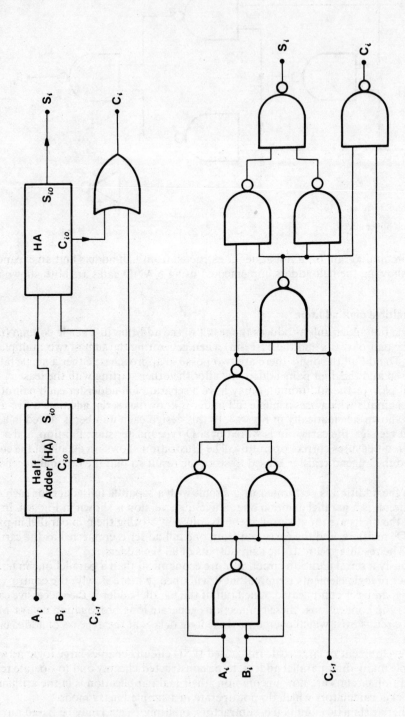

Fig. 8.2. Full-adder circuit

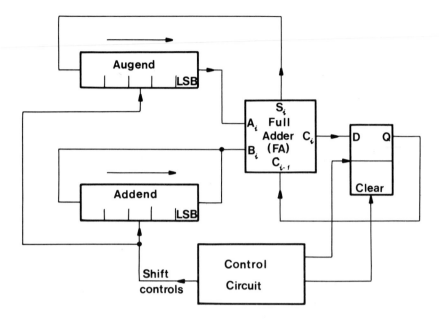

Fig. 8.3. Five-bit serial-addition circuit

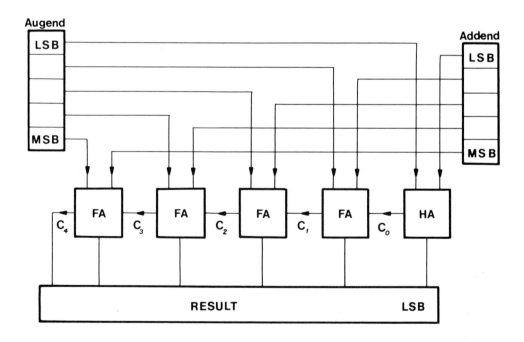

Fig. 8.4. Five-bit parallel-addition circuit

8.2 Counters

Only synchronous divide-by-N counters and asynchronous divide-by-2^n counters have been described in the earlier chapters. The general-purpose divide-by-N counter design method introduced in Chapter 6 is rather cumbersome when N is a large number. Provided that N has factors and that transient output states can be removed or tolerated a mixture of synchronous and asynchronous techniques may be used to construct counters with large values of N. For example a divide-by-one-thousand counter may be built by designing a divide-by-ten synchronous counter and connecting three of them in asynchronous series as in Fig. 8.5a. Similarly, a divide-by-sixty counter may be constructed from divide-by-ten and divide-by-six synchronous counters in asynchronous series as in Fig. 8.5b.

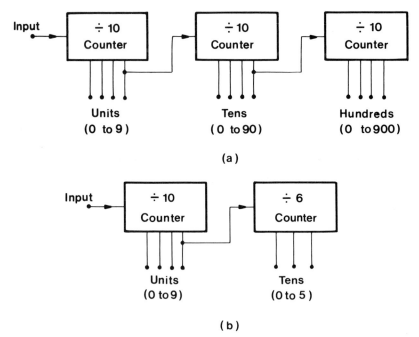

Fig. 8.5. Cascaded counter circuits

Many other methods of designing counters exist; several common techniques use serial shift registers and one of the most simple is the Johnson or 'twisted ring' counter. Assuming that the shift register is constructed from D-type master-slave flip-flops (other types may be used) a Johnson counter is formed by connecting the final stage $\bar{Q}$ output to the first-stage D input. The register is initially set so that all the bistable Q outputs are zero and the pulses to be counted are supplied to the register clock input. An n-stage Johnson counter will change through $2n$ different states; i.e. an n-stage Johnson counter is a divide-by-$2n$ counter. Figure 8.6 illustrates a five-stage Johnson counter circuit; Table 8.4 is its state table and illustrates the divide-by-ten action.

It is immediately apparent that a Johnson counter requires more bistables than a counter constructed using the design methods of Chapter 6. Also, since Johnson counters can only divide by $2n$, they cannot be designed to follow count sequences with an odd number of states. However this type of counter has the advantages that it is synchronous, requires no combinational logic circuits to supply flip-flop control inputs, and is simple to design. Some type of shift register based counter is usually chosen when a counter is

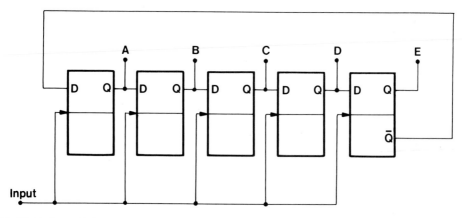

Fig. 8.6. Divide-by-ten Johnson counter

Table 8.4

| State | Shift register contents LSB ⟶ MSB | | | | | Input to register on next shift |
	A	B	C	D	E	$\bar{E}$
Initial, S_1	0	0	0	0	0	1
S_2	1	0	0	0	0	1
S_3	1	1	0	0	0	1
S_4	1	1	1	0	0	1
S_5	1	1	1	1	0	1
S_6	1	1	1	1	1	0
S_7	0	1	1	1	1	0
S_8	0	0	1	1	1	0
S_9	0	0	0	1	1	0
S_{10}	0	0	0	0	1	0
Initial, S_1	0	0	0	0	0	1

required as an internal element of an LSI circuit as such counters are the most compact in these applications.

Counter circuits are commonly used in digital measuring instruments; examples include point of sale weighing machines indicating weight and price, length measuring devices on machine tools, voltmeters, etc. These instruments usually operate by converting the quantity measured into a series of pulses whose number is proportional to the measurement (see Section 8.5). The pulses are input to a counter and the total count is displayed. In most applications the user does not wish to see the state of the counter while the count is in progress – only the final value is of interest.

The usual method of producing a display which only shows the total count is to use a display latch. This simply consists of a parallel in-parallel out (PIPO) register with the same number of stages as the counter. The outputs from all stages of the counter are connected as data inputs to the PIPO register, but while the count is in progress the clock input to the register is held so that no data is input. As soon as the count is complete, a clock pulse is applied to the PIPO register and the total count is then input to the register. The register outputs are connected to the display circuits which therefore always indicate the most recent total count.

8.3   Clock circuits

The terms 'clock' and 'clock circuit' have several meanings when applied to logic systems. They may refer to a circuit which indicates the actual time of day in hours, minutes and seconds as a conventional clock or watch does. Alternatively, a clock circuit may provide the timing control to some logic system and its outputs will consist of one or more chains of pulses at equal intervals in time (single or multi-phase clock).

In all cases the basic circuit construction is the same; a free-running oscillator supplies pulses at a constant frequency to a counter circuit and the counter outputs provide the indication of time. The oscillator frequency must be sufficiently high to provide output transitions at every instant required by the system.

One simple example of a clock circuit is that used in digital watches. This simply consists of a very accurate and stable quartz-crystal-controlled oscillator connected to counter and display circuits. The most commonly used oscillator frequency is 32 768 Hz(2^{15} Hz) and the first counter is a divide-by-32 768 one, which produces output pulses at one-second intervals. These one-second interval pulses are input to a divide-by-ten counter whose outputs may be used to display seconds and also provide the next stage input. This following stage is a divide-by-six counter providing the outputs for a tens of seconds display and the next stage input. The chain progresses, giving the appropriate outputs for minutes, tens of minutes, hours and date.

8.4   Memories

The development of LSI circuits has enabled large logic memory devices to be manufactured cheaply; these are now commonly-used components. A single bistable will store one binary digit (bit, logic state) and when the term memory is applied to a logic network it implies an organized arrangement of many storage elements. These storage elements are usually electronic bistable circuits although other two-state devices are sometimes used; some memory components for example are based on magnetic effects.

In a logic memory the storage elements are grouped into blocks of equal size; these blocks are called **words** and typical word lengths (sizes) are 1, 2, 4, 8 and 16 elements (bits). Each word is identified by a number called the **address** of the word; the first word is at address zero (not one), the second is at one, the third at two and so on. The most useful memories allow reference to any single word by application of logic levels representing the binary number for the address of the word at address control inputs. Such memories are termed random access memories; an important feature is that the time to refer to a particular word **(access time)** does not depend on the address. An alternative form of memory is a sequential (or cyclic or serial) access one in which the user must start at address zero and refer to each word in turn; in such a memory the access time for a particular word depends upon the address.

Memories may be divided into two further classes; they are either **read only memories** (ROMs) or **read and write memories**. In the case of a ROM, information has been permanently written into the memory in some way and cannot be changed; i.e. the user can only obtain (read) the contents of any address. There are several types of ROM and each has a different method of initial loading. Some are loaded by a process in the original manufacture while alternative types, called 'programmable read only memories' (PROMs), are loaded by the user during an irreversible process.

Those memories which allow the user to both read the contents of any location and enter new values (write) into any location are read and write memories but are known as RAMs (**random access memories**). This is confusing because most ROMs are also random access; normal use of ROM implies random access read only memory while RAM implies random access read and write memory.

In general a RAM is organized as shown schematically in Fig. 8.7. A ROM is arranged in a similar manner but there is no write (data input) facility. To enter information into a particular word of a RAM, logic values which represent the address of the word in binary form are applied to the address inputs. Simultaneously, the new data is supplied at the data inputs and a pulse is input to the write control. The contents of a word of a RAM or ROM are obtained by input of the address as in a write operation and application of a pulse to the read control; the contents of the word appear at the data outputs. Some designs of RAM use a tri-state bus system for the data and have a single set of data connections instead of separate input and output connections.

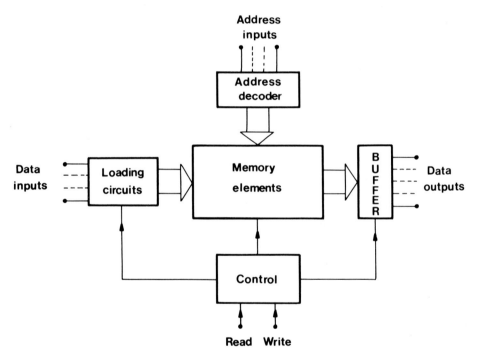

Fig. 8.7. Organization of a random access memory

Memories are available with several thousand storage elements in a single integrated circuit; one simple application of memories is to use a ROM in place of a complex multiple-output combinational logic circuit. When a ROM is used in this way, the address inputs are the combinational circuit inputs, the read control is arranged so that the ROM always outputs the contents of the currently selected address, and the data outputs are the circuit outputs. Circuit design simply consists of ensuring that the ROM contains the information which causes the circuit to behave as required by the truth table. Combinational circuits based on ROMs allow many complex multiple-output circuits to be constructed in a compact and inexpensive form; their main disadvantage is that they operate more slowly than circuits constructed from a network of basic logic gates.

8.5   Digital instruments

Many quantities formerly measured in an analogue manner using instruments with a
pointer moving across a scale are now measured in a digital form. The quantities
measured are inherently continuous (i.e. they are not in discrete units) and must be
converted into a discontinuous form for digital measurements. This conversion is
performed by converting the quantity into a chain of pulses and counting them, or
alternatively the quantity is converted into time and pulses from a clock circuit are
counted for this time period.

In most simple applications it is convenient to convert the quantity to be measured
into a voltage which is input to an electronic circuit that provides the digital output. This
electronic circuit is an **analogue to digital converter** (ADC), a circuit which performs the
reverse conversion is a **digital to analogue converter** (DAC).

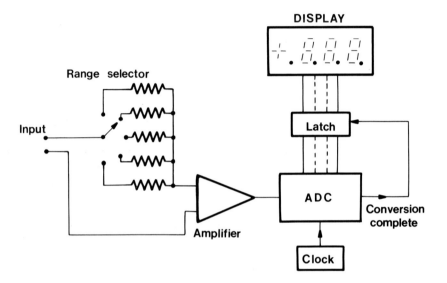

Fig. 8.8. A simple digital voltmeter

Both ADCs and DACs are available in integrated circuit form; the important
parameters for the designer incorporating them in a system are the accuracy of the
conversion and the time required to perform the conversion. A block diagram of a simple
multi-range digital voltmeter is shown in Fig. 8.8; in low-cost, moderate-quality
applications the complete circuit except for the range switch with its associated
components and the display may be incorporated in a single integrated circuit.

8.6   Comment

As integrated circuit manufacturing techniques develop, cheaper and more elaborate
devices will become available. These will reduce the design effort required in many
applications, but to ensure adequate production quantities the larger devices must be
multi-purpose ones. To use such components fully, designers will still require a wide
knowledge of logic network design techniques; only an introduction has been presented
here and an arbitrary selection of a few of the more advanced texts is given in the
bibliography.

Appendix A
The Binary number system

The number system which uses base two is the **binary system**, numbers in this system are written using only two symbols which represent a single unit (one) and nothing (zero). Usually the symbols chosen are 1 and 0 as they are already familiar from their everyday use in the decimal (denary) system. An example of a binary number is $101110011{\cdot}11001_2$ where the subscript 2 is used to indicate a binary number; using decimal notation this is

$$1 \times 2^7 + 0 \times 2^6 + 1 \times 2^5 + 1 \times 2^4 + 1 \times 2^3 + 0 \times 2^2 + 1 \times 2^1 + 1 \times 2^0 + 1 \times 2^{-1} + 1 \times 2^{-2}$$
$$+ 0 \times 2^{-3} + 0 \times 2^{-4} + 1 \times 2^{-5}$$

which is more conveniently written as

$$128 + 0 + 32 + 16 + 8 + 0 + 2 + 1 + 0{\cdot}5 + 0{\cdot}25 + 0 + 0{\cdot}031\,25 = 187{\cdot}781\,25$$

A.1 Conversion between binary and decimal systems

The most convenient method for conversion of a binary number into its equivalent in decimal form is the technique illustrated above in which each digit is multiplied by the appropriate value of 2^n expressed as a decimal number. The reverse conversion process from decimal to binary is not so obvious; it is most easily performed using the following technique which is a particular case of a general method for conversion between bases. In the conversion the decimal number is first separated into integral and fractional parts; these are then converted separately.

Conversion of the integral part is performed by dividing it by two, obtaining an integral result and noting the remainder. This process is repeated until the number is reduced to zero; the binary equivalent of the original integer is just the remainders in reverse order (i.e. the last remainder produced is the most significant binary digit). The following example shows the conversion of the decimal integer 363.

$$
\begin{array}{ll}
363 \div 2 = 181 & \text{remainder, } R = 1 \\
181 \div 2 = 90 & R = 1 \\
90 \div 2 = 45 & R = 0 \\
45 \div 2 = 22 & R = 1 \\
22 \div 2 = 11 & R = 0 \\
11 \div 2 = 5 & R = 1 \\
5 \div 2 = 2 & R = 1 \\
2 \div 2 = 1 & R = 0 \\
1 \div 2 = 0 & R = 1 \\
\end{array}
$$

The remainders in reverse order give 101101011 as the binary equivalent of 363 decimal; this result may be checked by the reverse conversion.

The fractional part of a decimal number is converted to binary form by repeated multiplication by two; the integer digit (1 or 0) produced to the left of the point is noted but neglected in subsequent multiplications. The integer digits are the required binary equivalent when written in the order in which they are produced (i.e. the first integer produced is the binary digit immediately to the right of the point). This process is illustrated by the conversion of the decimal fraction 0·413.

$$0 \cdot 413 \times 2 = 0 \cdot 826 \quad \text{integer, } I = 0$$
$$0 \cdot 826 \times 2 = 1 \cdot 652 \qquad\qquad I = 1$$
$$0 \cdot 652 \times 2 = 1 \cdot 304 \qquad\qquad I = 1$$
$$0 \cdot 304 \times 2 = 0 \cdot 608 \qquad\qquad I = 1$$
$$0 \cdot 608 \times 2 = 1 \cdot 216 \qquad\qquad I = 1$$
$$0 \cdot 216 \times 2 = 0 \cdot 432 \qquad\qquad I = 0$$
$$0 \cdot 432 \times 2 = 0 \cdot 864 \qquad\qquad 1 = 0$$
$$0 \cdot 864 \times 2 = 1 \cdot 728 \qquad\qquad I = 1$$
$$0 \cdot 728 \times 2 = 1 \cdot 456 \qquad\qquad I = 1$$
$$0 \cdot 456 \times 2 = 0 \cdot 921 \qquad\qquad I = 0$$
$$\text{etc.}$$

In general the conversion process continues indefinitely; a decimal fraction rarely corresponds to a finite length binary fraction. When the conversion is not exact it must be truncated at some stage; care must be taken to round the truncated number correctly. In the example, $0 \cdot 413$ decimal is equivalent to $0 \cdot 011010011$ binary to nine binary places.

Combining the two results gives that $363 \cdot 413$ decimal is equivalent to $101101011 \cdot 011010011$ binary.

A.2   Other conversions

Conversions between binary and octal (base eight) numbers or between binary and hexadecimal (base sixteen) are particularly simple as eight is 2^3 and sixteen is 2^4. An octal number will include only the digits 0, 1, 2, 3, 4, 5, 6 and 7; it is converted to binary simply by replacing each octal digit by the equivalent three-digit binary number. For the octal number $425 \cdot 713$ (often written $425 \cdot 713_8$) the conversion is

$$
\begin{array}{ccccccc}
4 & 2 & 5 & \cdot & 7 & 1 & 3 \\
100 & 010 & 101 & \cdot & 111 & 001 & 011
\end{array}
$$

giving $100010101 \cdot 111001011$ as the binary equivalent.

Reverse conversions are equally simple; starting from the point, a binary number is separated into groups of three digits and each group is converted into the equivalent octal digit. The conversion of $10111011 \cdot 11001_2$ into octal is illustrated below.

$$
\begin{array}{ccccc}
010 & 111 & 011 & \cdot & 110 & 010 \\
2 & 7 & 3 & \cdot & 6 & 2
\end{array}
$$

Thus $10111011 \cdot 11001_2$ is equivalent to $273 \cdot 62_8$; note the addition of zeros to complete the groups of three digits.

Hexadecimal to binary conversions and the reverse are equally simple and involve groups of four binary digits. Because conversions are so simple between the binary and octal or hexadecimal systems and because these systems are more compact than binary, they are often used in conjunction with computer systems.

Appendix B
Maxterm representation of circuits

In Chapter 3 a minterm representation was developed for combinational logic circuits and, as in all techniques involving Boolean variables, there is an equivalent dual method. This dual method is based on maxterms but, as two different definitions of a maxterm are in common use, there are two different maxterm representations of a circuit.

Definition 1 A **maxterm** is an OR function which includes every input variable (literal) once only in either true or complemented form. The maxterm can only have the value 0 when all the variables have values which are the inverse of those in the row of the truth table described by the maxterm.

Definition 2 A **maxterm** is that OR function which contains every literal once only in either true or complemented form. The OR function has the value 0 only if all the literals have the same values as those in the row of the truth table described by the maxterm.

Table B.1 lists all the possible values of three variables A, B and C with the corresponding maxterms according to both definitions.

Table B.1

C	B	A	Maxterm Definition 1	Definition 2
0	0	0	$\bar{A}+\bar{B}+\bar{C}$	$A+B+C$
0	0	1	$A+\bar{B}+\bar{C}$	$\bar{A}+B+C$
0	1	0	$\bar{A}+B+\bar{C}$	$A+\bar{B}+C$
0	1	1	$A+B+\bar{C}$	$\bar{A}+\bar{B}+C$
1	0	0	$\bar{A}+\bar{B}+C$	$A+B+\bar{C}$
1	0	1	$A+\bar{B}+C$	$\bar{A}+B+\bar{C}$
1	1	0	$\bar{A}+B+C$	$A+\bar{B}+\bar{C}$
1	1	1	$A+B+C$	$\bar{A}+\bar{B}+\bar{C}$

A detailed comparison of the two definitions is only relevant to those concerned with the development of methods for logic circuit design. Engineers applying such design methods will usually find that the techniques already described which use minterms are the most useful ones. There is a marginal advantage in using a maxterm approach when circuits constructed from NOR gates are required rather than circuits using NAND gates.

The two different maxterm definitions lead to two different methods of circuit design. Definition 1 is useful when generalized theorems of Boolean algebra are adopted but definition 2 is the more simple one to apply to the design of combinational logic circuits. A brief outline showing the development of a minimal expression for a logic circuit using maxterms based on definition 2 follows.

Table B.2 is the truth table of a three-input circuit which is described in minterms by

$$R = \bar{A}.\bar{B}.\bar{C}+\bar{A}.B.\bar{C}+\bar{A}.B.C$$

Table B.2

Inputs C B A	Output R	Maxterm (Definition 2)	Minterm
0 0 0	1	$A+B+C$	$\bar{A}.\bar{B}.\bar{C}$
0 0 1	0	$\bar{A}+B+C$	$A.\bar{B}.\bar{C}$
0 1 0	1	$A+\bar{B}+C$	$\bar{A}.B.\bar{C}$
0 1 1	0	$\bar{A}+\bar{B}+C$	$A.B.\bar{C}$
1 0 0	0	$A+B+\bar{C}$	$\bar{A}.\bar{B}.C$
1 0 1	0	$\bar{A}+B+\bar{C}$	$A.\bar{B}.C$
1 1 0	1	$A+\bar{B}+\bar{C}$	$\bar{A}.B.C$
1 1 1	0	$\bar{A}+\bar{B}+\bar{C}$	$A.B.C$

This is easily reduced to the sum of products form

$$R = \bar{A}.B + \bar{A}.\bar{C}$$

which can be expressed in NAND functions as

$$R = \overline{(\overline{\bar{A}.B}).(\overline{\bar{A}.\bar{C}})}.$$

Using definition 2 the maxterms for which the circuit output, R, is 0 are $\bar{A}+B+C$, $\bar{A}+\bar{B}+C$, $A+B+\bar{C}$, $\bar{A}+B+\bar{C}$ and $\bar{A}+\bar{B}+\bar{C}$. Each maxterm in this list can only be 0 for input conditions corresponding to the values of the inputs in the row of the truth table identified by the maxterm. As a 0 dominates an AND function and R is 1 unless one of the listed maxterms is 0 then R is the AND function of all these maxterms; i.e.

$$R = (\bar{A}+B+C).(\bar{A}+\bar{B}+C).(A+B+\bar{C}).(\bar{A}+B+\bar{C}).(\bar{A}+\bar{B}+\bar{C})$$

It is simple to develop a Karnaugh map with the squares identified by maxterms. A three-variable map is shown in Fig. B.1a and the completed map for the example is Fig. B.1b. As the method is a dual of the previous minterm technique, groups of zeros are formed on the map rather than groups of ones. The variables which change within a group are eliminated and the groups are described by sum terms which are combined in a product expression. Two groups are shown in Fig. B.1b, the group of four is simply $\bar{A}$ and the group of two is $B+\bar{C}$; these combine giving

$$R = \bar{A}.(B+\bar{C})$$

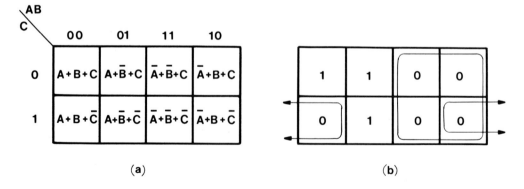

(a)   (b)

Fig. B.1.

This result is a product of sums expression and it is easily converted into the sum of products result obtained using minterms. Alternatively, double inversion and application of de Morgan's theorem leads to the NOR function form

$$R = A + (\overline{\overline{B} + \overline{C}}).$$

Bibliography

The following list is a selection of books which are concerned with logic circuits and digital computers; most of the books are more advanced than this one. Several of the books at the end of the list are published by integrated circuit manufacturers and usually they can only be obtained from the manufacturers or their appointed distributors.

Bannister, B. R. and Whitehead, D. G., *Fundamentals of Digital Systems*, McGraw-Hill, 1973.

Booth, T. L., *Digital Networks and Computer Systems*, Wiley and Sons, 1971.

Cripps, M., *An Introduction of Computer Hardware*, Edward Arnold, 1977.

Friedman, A. D., *Logical Design of Digital Systems*, Pitman, 1977.

Hill, F. J. and Peterson, G. R., *Introduction to Switching Theory and Logical Design* (2nd Edition), Wiley and Sons, 1974.

Lewin, D., *Logical Design of Switching Circuits* (2nd Edition), Nelson, 1974.

Lewin, D., *Theory and Design of Digital Computers*, Nelson, 1972.

Lind, L. F. and Nelson, J. C. C., *Analysis and Design of Sequential Digital Systems*, MacMillan, 1977.

Mano, M. M., *Computer Logic Design*, Prentice-Hall Inc., 1972.

Morris, N. M., *Logic Circuits* (2nd Edition), McGraw-Hill, 1976.

Morris, R. L. and Miller, J. R. (Eds), *Designing with TTL Integrated Circuits*, McGraw-Hill, 1971.

Mowle, F. J., *A Systematic Approach to Digital Logic Design*, Addison-Wesley, 1976.

Nashelsky, L., *Introduction to Digital Computer Technology*, Wiley and Sons, 1972.

Peatman, J. B., *The Design of Digital Systems*, McGraw-Hill, 1972.

Tocci, R. J., *Digital Systems; Principles and Applications*, Prentice-Hall Inc., 1977.

Townsend, R., *Digital Computer Structure and Design*, Newnes-Butterworths, 1975.

Woollons, D. J., *Introduction to Digital Computer Design*, McGraw-Hill, 1972.

Norris, B. (Ed.), *Semiconductor Circuit Design* (*Volume* 2), Texas Instruments Limited (U.K.), 1972.

Mullard, *Mullard TTL Integrated Circuits. Applications* (2nd Edition), Mullard Limited.

Motorola, *McMOS Handbook*, Motorola Inc., 1972.

R.C.A., *COS/MOS Integrated Circuits Manual*, R.C.A. Corp., 1972.

Index